Applications of Vector Analysis and Complex Variables in Engineering

Otto D. L. Strack

Applications of Vector Analysis and Complex Variables in Engineering

Springer

Otto D. L. Strack
CEGE Department
University of Minnesota
MINNEAPOLIS, MN, USA

ISBN 978-3-030-41170-1 ISBN 978-3-030-41168-8 (eBook)
https://doi.org/10.1007/978-3-030-41168-8

This Springer imprint is published by the registered company Springer Nature Switzerland AG
The registered company address is: Gewerbestrasse 11, 6330 Cham, Switzerland

To Andrine

Contents

Preface

The primary objective of this book is to teach the application of mathematics to engineering problems. We do not emphasize the proofs of the various approaches and theorems that we will use. This in no way means that mathematical proofs are not important; without them mathematics would not exist as a useful tool. However, although we must be very careful in our applications of mathematical methods and theorems to solve our problems, we will take certain proofs (uniqueness proofs are an example) for granted.

We cover vectors in three-dimensional space in Chapter 1; rather than using the common symbolic notation of vectors and tensors, we adopt the Einstein summation convention and the strict application of indicial notation, closely following the approach by Duschek and Hochreiner, presented in *Tensorrechnung in Analytischer Darstellung: I Tensoralgebra (5th edition, 1968), II Tensoranalysis (2nd edition, 1961), and III Anwendungen in Physik und Technik, 2nd edition, 1965)* by A. Duschek and A. Hochreiner, Wien, Springer Verlag (in German).

We cover vector fields in Chapter 2, explaining the Lagrangian and Eulerian descriptions. We cover applications of vectors and vector fields to fluid mechanics in Chapter 3, using this application to introduce divergence and curl to the reader, and cover flow with both divergence and rotation (the general field). We discuss integral theorems in Chapter 4. We introduce coordinate transformations with straight axes in Chapter 5, using statics of solids to introduce the reader to tensors of the second rank, the gradient, invariants, stresses, displacements and strains, and the compatibility conditions.

We cover partial differential equations of the first order in Chapter 6, illustrated by an application to contaminant transport. Linear partial differential equations of the second order are the subject of Chapter 7, where we distinguish the three types: elliptic, parabolic, and hyperbolic. We present the characteristic equations that define the type of any second order linear partial differential equation. We cover elliptic partial differential equations using complex variables for solving two-dimensional elliptic problems in Chapter 8. We use the complex variable representation for explaining the Helmholtz decomposition theorem, and integral theorems. We cover applications of complex variables to fluid mechanics, groundwater flow, and linear elasticity in Chapter 9. We devote Chapter 10 to parabolic partial differential equations, where we present several solutions of the diffusion equation, taken from heat flow, soil mechanics, and groundwater flow. We introduce and apply the Laplace transform and separation of variables to obtain these solutions. We cover the hyperbolic case in Chapter 11, with longitudinal and transverse vibrations as examples. We introduce the quasi-linear hyperbolic partial differential equation in Chapter 12, with failure of a granular medium taken as the example. The final Chapter, Chapter 13, is devoted to the Navier-Stokes equation.

Acknowledgements

This textbook could not have been completed without the help of my wife Andrine, who found many errors in both the text and the equations after repeated reading. She typed substantial parts of the text and made the corrections.

The book came about during years of teaching mathematics to civil engineering students, who made numerous useful comments. I am indebted to the reviewers; I thank them for their comments and suggestions for improvement.

Chapter 1

Vectors in Three-Dimensional Space

We introduce vector analysis using fluid mechanics as the vehicle for providing physical meaning to the concepts of vectors and the associated definitions and operations.

A vector is a quantity that has both a length and an orientation. A description of a vector in three dimensions requires three parameters, for example the components of the vector in the three coordinate directions, or the length of the vector and two angles,.

We use vectors to represent physical quantities, such as the velocity of a fluid particle. The orientation and magnitude of vectors are independent of the choice of the coordinate system; if the coordinate system changes, the components of vectors must change in such a way that both the length and orientation of the vectors remain the same.

We represent the vectors by their components, that is, by their projections onto the three Cartesian coordinate axes. If the velocity of a water particle has a magnitude v, we write the three components as v_x, v_y, v_z or as v_1, v_2, v_3:

$$v^2 = v_1^2 + v_2^2 + v_3^2 = \sum_{i=1}^{3} v_i v_i. \tag{1.1}$$

We use a compact notation introduced by Einstein, the Einstein Summation Convention: summation is implicit when an index occurs twice (and not more than twice) in a product:

$$v^2 = v_1^2 + v_2^2 + v_3^2 = v_i v_i. \tag{1.2}$$

We obtain further compactness by representing a vector by its components not as (v_1, v_2, v_3), but simply as v_i, where i is the index, which represents all three components of the vector. Such an index is called a free index, it is free to represent either 1, 2, or 3. The free index is in contrast with the index i in (1.2), which is called a dummy index; it cannot be chosen to represent either 1, 2, or 3 but is summed over.

O. D. L. Strack, *Applications of Vector Analysis and Complex Variables in Engineering*, https://doi.org/10.1007/978-3-030-41168-8_1

1.1 Unit Vectors

As seen in (1.1), we denote the magnitude, or modulus, of a vector by omitting its index: $v^2 = v_i v_i$. Unit vectors are special vectors that have a length 1. Thus, if T_i represents a unit vector, then, by (1.2),

$$T^2 = 1 = T_i T_i. \tag{1.3}$$

The vector T_i has only two degrees of freedom; (1.3) represents a relationship between the three components of T_i. Each of the components T_i is the cosine of the angle between T_i and a coordinate axis; the components T_i are known as the direction cosines of the vector.

We may write a vector as the product of its magnitude and a unit vector that represents its orientation; if the velocity of a water particle is v and it is oriented in the direction of a unit vector T_i, then:

$$v_i = vT_i, \tag{1.4}$$

see Figure 1.1.

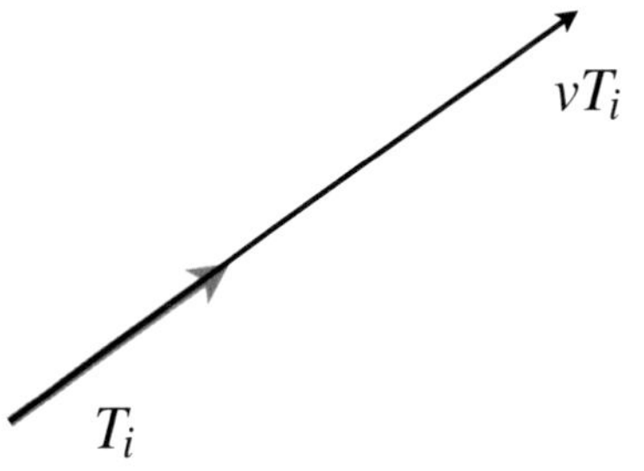

Figure 1.1: A unit vector to define direction

1.2 Comparison with Symbolic Notation

Indicial notation is different in principle from symbolic notation. A vector in symbolic notation could be represented as $\vec{A}$. If we introduce base vectors $\vec{i}, \vec{j}, \vec{k}$ and the components of $\vec{A}$ as A_i, A_j, A_k, then the vector is represented as

$$\vec{A} = A_i\vec{i} + A_j\vec{j} + A_k\vec{k}. \tag{1.5}$$

In indicial notation, we label the coordinate axes with numbers, $1, 2, 3$, rather than with symbols, e.g., x, y, z, and represent the coordinates of a point as x_i, where i represents the coordinate in question. Thus, instead of representing the vector in the form (1.5), we represent it by its components A_i in the three coordinate directions. This is illustrated in Figure 1.2. We present some of the equations in symbolic form behind the one in indicial notation.

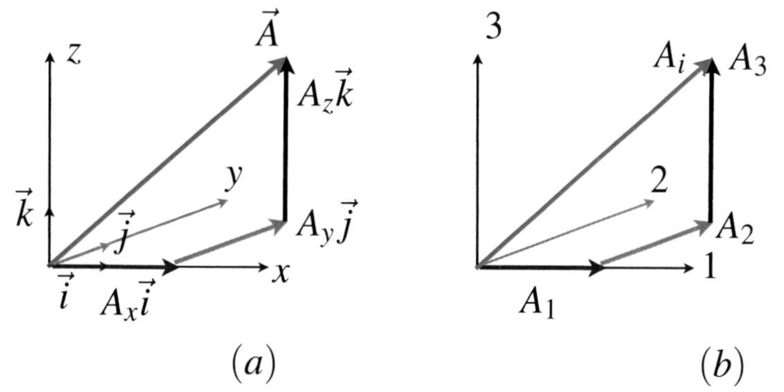

Figure 1.2: Symbolic (a) versus indicial (b) notation

1.3 The Dot Product

Let A_i, B_i, and C_i represent three vectors that form the three sides of a triangle:

$$C_i = A_i - B_i \qquad \vec{C} = \vec{A} - \vec{B}. \tag{1.6}$$

We apply the law of cosines to the triangle as illustrated in Figure 1.3, with α the angle between A_i and B_i:

$$C^2 = A^2 + B^2 - 2AB\cos\alpha. \tag{1.7}$$

Recall that we represent the modulus (length) of a vector by using the symbol for the vector without its index. We obtain from (1.6):

$$C^2 = C_iC_i = (A_i - B_i)(A_i - B_i) = A_iA_i - 2A_iB_i + B_iB_i = A^2 + B^2 - 2AB\cos\alpha, \tag{1.8}$$

so that, with $A_iA_i = A^2$ and $B_iB_i = B^2$:

$$A_iB_i = AB\cos\alpha \qquad \vec{A}\cdot\vec{B} = AB\cos\alpha. \tag{1.9}$$

1.3.1 Projection of a Vector onto a Given Direction

We obtain an expression for the projection of a vector v_i onto a line with orientation defined by the unit vector T_i as the product of the magnitude of the vector, v, and the unit vector T_i:

$$\underset{t}{v} = v_iT_i = v_1T_1 + v_2T_2 + v_3T_3 \qquad \underset{t}{v} = v\cos\alpha \qquad \vec{v}\cdot\vec{T}, \tag{1.10}$$

where α is the angle between v_i and the unit vector T_i. This is the special case of the dot product of two vectors, say A_i and B_i, where B_i is a unit vector. This is illustrated in Figure 1.4.

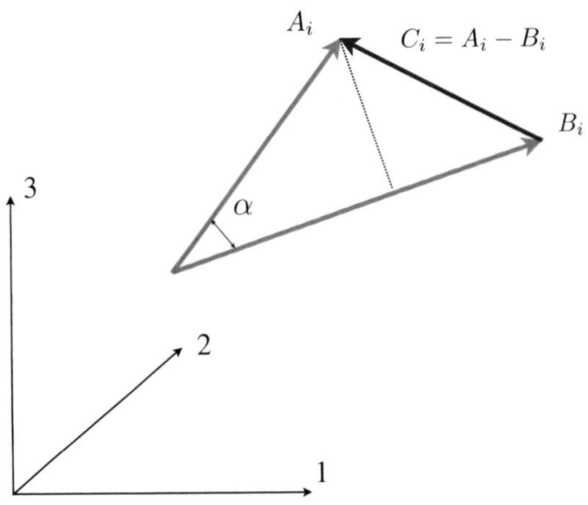

Figure 1.3: Vectorial addition and the dot product

1.4 The Cross Product

The cross product of two vectors is a vector that is perpendicular to the two vectors in the cross product. We write the cross product with the help of a quantity called the alternating ε-tensor of the third rank, ε_{ijk}, or the Levi-Civita symbol. We define a tensor formally in Chapter 5, but for now merely state that it is a quantity that carries more than one index, as opposed to a vector, which has only one. The ε tensor is defined as follows: $\varepsilon_{ijk} = 1$ for an odd permutation of 123, which is an even permutation of $ijk = 123$ and $\varepsilon_{ijk} = -1$ for an even permutation of $ijk = 132$; we obtain an even permutation by shifting the numbers past the right end of the string, placing at the front the number that we dropped at the end. Thus,

$$\begin{aligned} \varepsilon_{123} = \varepsilon_{312} = \varepsilon_{231} &= 1 \\ \varepsilon_{132} = \varepsilon_{213} = \varepsilon_{321} &= -1. \end{aligned} \tag{1.11}$$

The value of ε_{ijk} is zero when any two of the three indices are equal. Interchanging the position of any two of the indices results in a sign change.

We write the cross product of the vectors A_i and B_i, as a vector C_i:

$$C_i = \varepsilon_{ijk} A_j B_k \qquad \vec{C} = \vec{A} \times \vec{B}. \tag{1.12}$$

The index i in this equation represents either 1, 2, or 3, whereas the indices j and k are to be summed over, according to the Einstein summation convention. The index i is the free index. The summed-over indices j and k can be given any name; these indices are dummy indices. Thus, (1.12) is identical to the following equation:

$$C_i = \varepsilon_{ipq} A_p B_q. \tag{1.13}$$

We emphasize the distinction between *free* indices and *dummy* indices. A free index represents any number between 1 and 3. Dummy indices are summed over. Free

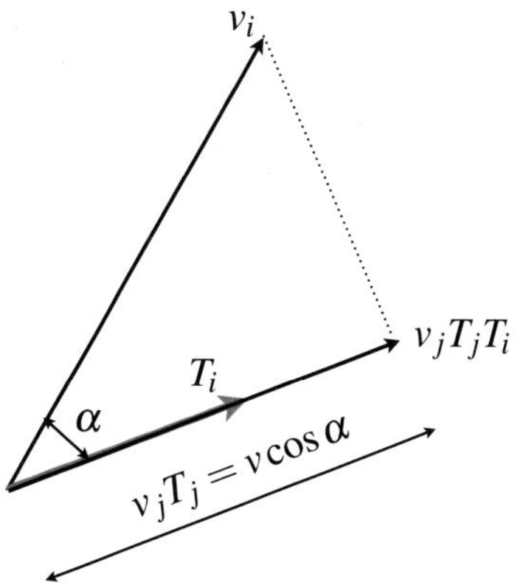

Figure 1.4: Projection of a vector onto a unit vector of given direction

indices can only appear once in each term, and must match on the two sides of an equation. Thus, if the free index is m to the left of the equal sign, it must be m to the right of the equal sign. Dummy indices must appear in pairs, but never more than in a single pair per term, and can be any letter.

We see that C_i is perpendicular to A_i, see Figure 1.5, by forming the dot product of C_i and A_i. If these two vectors are indeed orthogonal, then their dot product must be zero, because the projection of one vector onto another that it is perpendicular to it is zero. It follows that

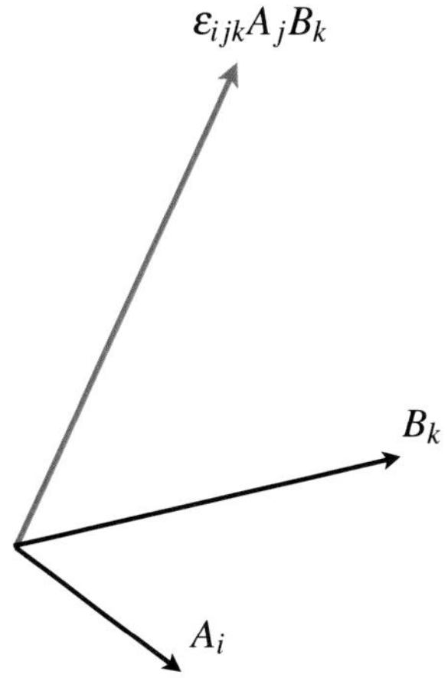

Figure 1.5: The cross product

$$C_iA_i = \varepsilon_{ijk}A_iA_jB_k \qquad \vec{C}\cdot\vec{A} = (\vec{A}\times\vec{B})\cdot\vec{A} \tag{1.14}$$

The order of the vectors is immaterial; the procedure is uniquely defined by the Einstein summation convention and the letters of the indices. We may interchange the indices in (1.14) of the two vectors A_i and A_j. Thus,

$$C_iA_i = \varepsilon_{ijk}A_iA_jB_k = \varepsilon_{ijk}A_jA_iB_k, \tag{1.15}$$

or, renaming the dummy index i to j and j to i,

$$C_iA_i = \varepsilon_{ijk}A_iA_jB_k = \varepsilon_{jik}A_iA_jB_k. \tag{1.16}$$

Because $\varepsilon_{ijk} = -\varepsilon_{jik}$, this becomes

$$C_iA_i = \varepsilon_{ijk}A_iA_jB_k = -\varepsilon_{ijk}A_iA_jB_k = 0 \qquad (\vec{A} \times \vec{B}) \cdot \vec{A} = 0, \tag{1.17}$$

where we use that a quantity that is both positive and negative must be zero. Thus, C_i is indeed perpendicular to both the vectors A_i and B_i.

We deduce from (1.17) the general rule that the cross product of two parallel vectors is zero. *We see that when ε_{ijk} operates on a tensor that is symmetrical in any two of the indices i, j, k, the result is zero.*

We derive an expression for the magnitude of C_i by writing

$$C_iC_i = C^2 = \varepsilon_{ipq}C_iA_pB_q = \varepsilon_{imn}A_mB_n\varepsilon_{ipq}A_pB_q = \varepsilon_{imn}\varepsilon_{ipq}A_mB_nA_pB_q. \tag{1.18}$$

This equation in symbolic notation is

$$(\vec{A} \times \vec{B}) \cdot (\vec{A} \times \vec{B}). \tag{1.19}$$

We demonstrate in what follows that this can be written as

$$C_iC_i = C^2 = A_mA_mB_nB_n - (A_nB_n)^2 = A^2B^2 - (AB)^2\cos^2\alpha, \tag{1.20}$$

or

$$C^2 = (AB)^2(1 - \cos^2\alpha) = (AB\sin\alpha)^2, \tag{1.21}$$

where A and B are the magnitudes of the vectors A_i and B_i.

$$A = \sqrt{A_iA_i} \quad B = \sqrt{B_iB_i}. \tag{1.22}$$

Expression (1.21) is the formula for the square of the area of the parallelogram enclosed by the vectors A_i and B_i. The cross product is a convenient way to obtain a vector that is perpendicular to two other vectors; if A_i and B_i are vectors in the plane, then the unit vector normal to the plane is n_i with

$$n_i = \frac{\varepsilon_{ijk}A_jB_k}{C}. \tag{1.23}$$

Problem 1.1

Consider equation (1.18)

$$C_iC_i = C^2 = \varepsilon_{imn}\varepsilon_{ipq}A_mB_nA_pB_q,$$

and demonstrate that (1.20) is correct by letting i be 1, 2, and 3 in turn, writing out the implicit summation and combining terms.

1.5 Free Indices and Dummy Indices

The dummy indices are always indices that appear in pairs in an expression and are to be summed over. The dummy indices can only occur as a single pair in each term in an expression. Free indices represent the numbers 1, 2, or 3; their value can be chosen to represent any one of these three numbers, hence the name 'free index'. Free indices must match on the two sides of an equation, and must appear in each term, and always singly. It is important, when using indicial notation, to verify that each equation matches these fundamental rules of indicial notation.

1.6 Base Vectors

A vector in symbolic notation is represented in terms of the three base vectors $\vec{i},\vec{j},\vec{k}$, as is illustrated in Figure 1.2. The three components of $\vec{A}$ are often represented in symbolic notation as A_i, A_j, A_k, which tends to cause some adjusting when comparing symbolic notation with indicial notation. The vector $\vec{A}$ thus would be written as

$$\vec{A} = A_i\vec{i} + A_j\vec{j} + A_k\vec{k}. \tag{1.24}$$

1.7 The Kronecker Delta

The Kronecker delta is represented as δ_{ij} and is a tensor of the second rank. The Kronecker delta is defined such that $\delta_{ij} = 1$ for $i = j$ and $\delta_{ij} = 0$ for $i \neq j$.

The Kronecker delta is a tensor, but can be used to represent the three base vectors in a Cartesian coordinate system. There is a connection between the three base vectors $\vec{i},\vec{j},\vec{k}$ in (1.5)

$$\begin{aligned} \vec{i} &\equiv \delta_{1i} = (1,0,0) \\ \vec{j} &\equiv \delta_{2i} = (0,1,0) \\ \vec{k} &\equiv \delta_{3i} = (0,0,1). \end{aligned} \tag{1.25}$$

We can represent the vector with components A_i in a manner similar to (1.5):

$$\vec{A} \equiv A_1\delta_{1i} + A_2\delta_{2i} + A_3\delta_{3i} = A_k\delta_{ki} = A_i. \tag{1.26}$$

Multiplying any quantity, such as a vector A_i, by the Kronecker delta results in modification of the index of the vector:

$$A_j = \delta_{ij}A_i. \tag{1.27}$$

This is true because the only term in the product $\delta_{ij}A_i$ that is not zero is the one where $i = j$; all the other terms are zero because the Kronecker delta is zero unless its two indices are equal.

Problem 1.2

Apply the following equation to demonstrate that this yields the equation given in problem 1.1

$$\delta_{ip}\delta_{jq} - \delta_{iq}\delta_{jp} = \varepsilon_{ijk}\varepsilon_{kpq}.$$

Question:
Is the result of problem 1.1 sufficient to prove that the relationship is correct?

1.8 The Levi-Civita Symbol and the Kronecker Delta

We use the following relationship between ε tensor and Kronecker delta:

$$\boxed{\varepsilon_{ijk}\varepsilon_{imn} = \delta_{jm}\delta_{kn} - \delta_{jn}\delta_{km}.} \tag{1.28}$$

We verify this equation by summing over i, and then substituting values for j,k,m and n. This is a tedious and lengthy process, but will prove the validity of (1.28). One of the advantages of indicial notation (sometimes called engineering notation) is that all equations can be verified, and that a few basic rules are sufficient to derive with relative ease many equations that otherwise would have to be remembered or looked up.

We verify the validity of (1.28) without writing out all possible combinations of the four free indices j,k,m,n as follows. We first sum over the index i

$$\varepsilon_{ijk}\varepsilon_{imn} = \varepsilon_{1jk}\varepsilon_{1mn} + \varepsilon_{2jk}\varepsilon_{2mn} + \varepsilon_{3jk}\varepsilon_{3mn}. \tag{1.29}$$

The expression $\varepsilon_{ijk}\varepsilon_{imn}$ is zero if $j = k$ or if $m = n$, i.e.,

$$\varepsilon_{ijk}\varepsilon_{imn} = 0 \qquad j = k \tag{1.30}$$

$$\varepsilon_{ijk}\varepsilon_{imn} = 0 \qquad m = n. \tag{1.31}$$

If both $j \neq k$ and $m \neq n$, only one term in the sum (1.29) is non-zero. For example, if $j = 1$ and $k = 2$, only the third term is non-zero, if either $m = 1, n = 2$, or $m = 2, n = 1$. In the former case the expression in (1.29) is 1, and in the latter case, it is -1. We repeat this process for all possible cases $j \neq k, m \neq n$, and obtain:

$$\varepsilon_{ijk}\varepsilon_{imn} = 1 \qquad j = m \qquad k = n \tag{1.32}$$

$$\varepsilon_{ijk}\varepsilon_{imn} = 0 \qquad j = m \qquad k \neq n \tag{1.33}$$

$$\varepsilon_{ijk}\varepsilon_{imn} = -1 \qquad j = n \qquad k = m \tag{1.34}$$

$$\varepsilon_{ijk}\varepsilon_{imn} = 0 \qquad j = n \qquad k \neq m. \tag{1.35}$$

The behavior represented by (1.29) through (1.35) matches the expression:

$$\varepsilon_{ijk}\varepsilon_{imn} = \delta_{jm}\delta_{kn} - \delta_{jn}\delta_{km}. \tag{1.36}$$

We use (1.28) to derive laws and properties used in fluid mechanics. The latter identity also lets us obtain the result reported in (1.20); application of this equation to the product of the alternating tensors in (1.18) directly leads to (1.20).

Problem 1.3
Use indicial notation to demonstrate that

$$(\vec{A} \times \vec{B}) \cdot (\vec{C} \times \vec{D}) = (\vec{A} \cdot \vec{C})(\vec{B} \cdot \vec{D}) - (\vec{A} \cdot \vec{D})(\vec{B} \cdot \vec{C}).$$

Hint: use the relationship $\varepsilon_{ijk}\varepsilon_{pqk} = \delta_{ip}\delta_{jq} - \delta_{iq}\delta_{jp}$.

Problem 1.4

Consider three vectors, a_i, b_i, and c_i, where

$$a_i = \lambda b_i + \mu c_i,$$

where λ and μ are constants. Work out the following expression

$$d = a_i \varepsilon_{ijk} b_j c_k.$$

Problem 1.5

Demonstrate that the following equation holds:

$$\varepsilon_{ijk} b_j c_k = -\varepsilon_{ijk} b_k c_j.$$

Problem 1.6

Explain, using indicial notation and the meaning of the various implicit summations, that the following expression:

$$V = \varepsilon_{ijk} a_i b_j c_k,$$

is equal to the volume of the parallelepiped formed by the three vectors a_i, b_i, and c_i.

Problem 1.7

Consider the following expression for an arbitrary point x_i on a line $\mathscr{L}$ defined by

$$x_i = \overset{0}{x}_i + \lambda a_i,$$

where λ is an arbitrary scalar and a_i a vector. Next consider the plane $\mathscr{P}$ defined by $\mathscr{L}$ and a given point $\overset{1}{x}_i$. The vector v_i represents a velocity.

Questions:

1. Determine an expression for the component of v_i normal to the plane $\mathscr{P}$.
2. Determine an expression for the component of v_i tangential to the plane $\mathscr{P}$
3. Answer the preceding two questions for the case that: $\overset{0}{x}_i = (0,1,5)$, a_i=(1,2,-3), $\overset{1}{x}_i$=(6,2,1), and v_i=(-2,3,4).

Problem 1.8

Demonstrate that

$$\mathbf{A} \times (\mathbf{B} \times \mathbf{C}) + \mathbf{B} \times (\mathbf{C} \times \mathbf{A}) + \mathbf{C} \times (\mathbf{A} \times \mathbf{B}) = \mathbf{0}.$$

Problem 1.9

Demonstrate that

$$(\mathbf{A} \times \mathbf{B}) \cdot (\mathbf{B} \times \mathbf{C}) \times (\mathbf{C} \times \mathbf{A}) = [\mathbf{ABC}]^2,$$

where

$$[\mathbf{ABC}] = \varepsilon_{ijk} A_i B_j C_k.$$

Problem 1.10

Consider the following three vectors

$$\mathbf{U} = \frac{\mathbf{B} \times \mathbf{C}}{[\mathbf{ABC}]} \qquad \mathbf{V} = \frac{\mathbf{C} \times \mathbf{A}}{[\mathbf{ABC}]} \qquad \mathbf{W} = \frac{\mathbf{A} \times \mathbf{B}}{[\mathbf{ABC}]}.$$

Show that

$$\mathbf{A} \cdot \mathbf{U} = \mathbf{B} \cdot \mathbf{V} = \mathbf{C} \cdot \mathbf{W} = 1.$$

Chapter 2

Vector Fields

The vectors that we deal with in engineering usually are neither constant in space, nor in time. Such vectors are functions of position, i.e., they have different values depending on where we look; the collection of such vectors are called vector fields. We are dealing with a time-varying vector field if the vectors vary with time. An example of a time-varying vector field is the velocity field that represents the flow of particles in the air around us; another example is the vector field that represents the flow of water in a river. Different velocity values are found at different locations, and at different times.

2.1 Field Lines: Path Lines and Stream Lines

The water particles follow some curvilinear paths through time and space; these curves are path lines. We examine the properties of path lines from both the Lagrangian and Eulerian viewpoints. The Lagrangian view is to move with the particle through time and space. The Eulerian view is to be at a fixed point in space, and consider the velocities of the particles over time as they move by.

2.1.1 The Lagrangian Description

We consider a particle moving through three-dimensional space over some path; we represent the coordinates of the particle as x_i. The coordinates are functions of time only, given a specific particle:

$$x_i = x_i(t). \tag{2.1}$$

We denote the distance traveled by the particle along its path line by s; s is a function of time:

$$s = s(t). \tag{2.2}$$

The velocity of the particle is the derivative of s with respect to time:

$$v = \frac{ds}{dt} = \dot{s}. \tag{2.3}$$

O. D. L. Strack, *Applications of Vector Analysis and Complex Variables in Engineering*, https://doi.org/10.1007/978-3-030-41168-8_2

The dot above the symbol indicates differentiation with respect to time. The three components of the velocity vector in space are the derivatives of the three coordinates of the particle with respect to time:

$$v_i = \frac{dx_i}{dt} = \dot{x}_i. \tag{2.4}$$

The derivative of the coordinates x_i with respect to s is the tangent unit vector to the path line:

$$T_i = \frac{dx_i}{ds}, \tag{2.5}$$

where T_i is a unit vector, see Figure 2.1,

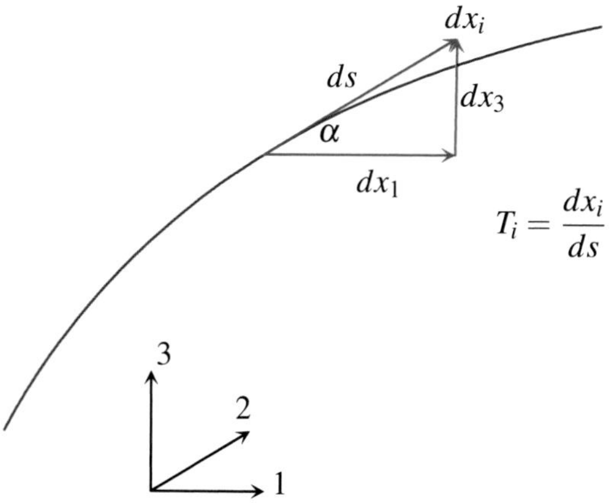

Figure 2.1: The tangent vector T_i to a curve

$$T_iT_i = \frac{dx_idx_i}{(ds)^2} = \frac{dx_1^2 + dx_2^2 + dx_3^2}{(ds)^2} = 1. \tag{2.6}$$

The acceleration of the particle is the rate of change of the velocity, i.e.,

$$a_i = \dot{v}_i = \ddot{x}_i \tag{2.7}$$

It follows from the chain rule and (2.4) that:

$$v_i = \frac{dx_i}{dt} = \frac{dx_i}{ds}\frac{ds}{dt} = vT_i. \tag{2.8}$$

We substitute this expression for v_i in (2.7):

$$a_i = \frac{d(vT_i)}{dt} = \frac{dv}{dt}T_i + v\frac{dT_i}{dt} = \frac{dv}{dt}T_i + v\frac{dT_i}{ds}\frac{ds}{dt}, \tag{2.9}$$

or

$$a_i = \frac{dv}{dt}T_i + v^2\frac{dT_i}{ds}. \tag{2.10}$$

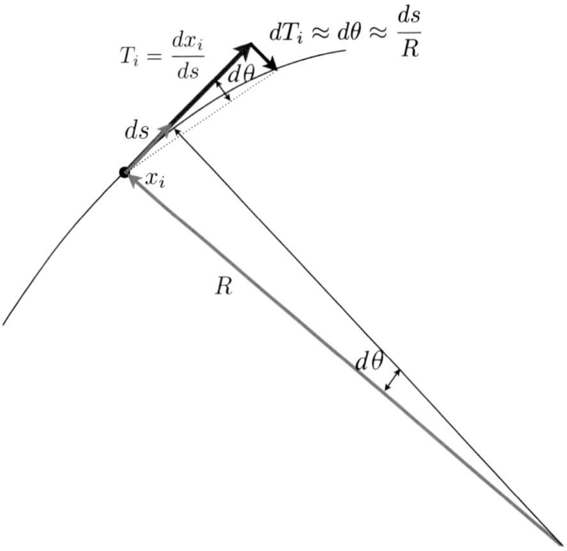

Figure 2.2: Tangent vector and its derivative along the pathline

The vector T_i is a unit vector; its length cannot change, so that the incremental change of T_i, dT_i, is perpendicular to T_i. If we move in the direction of T_i over a distance ds, then $ds = Rd\theta$ where $d\theta$ is the change in angle of T_i, and R is the radius of the circle that touches the curve and shares its curvature at the point of touching (see Figure 2.2, where R is the radius of curvature). The unit vector T_i rotates over $d\theta$, the magnitude of its change is $d\theta$:

$$\frac{dT_i}{ds} = \frac{d\theta}{ds}N_i = \frac{d\theta}{Rd\theta}N_i = \frac{1}{R}N_i, \tag{2.11}$$

where N_i is the unit vector normal to the curve, the unit normal that points in the direction of the change of T_i. We use (2.11) to write (2.10) as:

$$a_i = \frac{dv}{dt}T_i + \frac{v^2}{R}N_i = \underset{T}{a}T_i + \underset{N}{a}N_i, \tag{2.12}$$

where $\underset{T}{a}$ is the component of the acceleration tangent to the path line, and $\underset{N}{a}$ is the component of acceleration normal to the path line; the latter represents the centripetal acceleration.

2.1.2 The Eulerian Description

The Lagrangian description is commonly used in the mechanics of masses moving through space; it is less popular in fluid mechanics, where the Eulerian view is preferred. Although the results are demonstrably the same, regardless of the approach used for the derivation, the analyses differ considerably and the forms of the results differ as well.

In the Eulerian view, we represent the velocities of the particles as a time-varying vector field; the velocity is a vector that is a function of both space and time:

$$v_i = v_i(x_1, x_2, x_3, t) = v_i(x_k, t), \tag{2.13}$$

where we replaced (x_1, x_2, x_3) by x_k for convenience, being careful to chose a different index than i to avoid confusion (the same index could be interpreted as implying summation).

We cannot obtain an expression for the acceleration simply by differentiating (2.13) partially with respect to time; the particles move through space as well as through time. This differs from the Lagrangian viewpoint, where the velocity is attached to the particle, and is a function of time only; the movement was implicitly represented by defining the velocity as the rate of change with time of the coordinates of the particle.

We need to differentiate (2.13) in a special way in order to determine the acceleration. As an introduction, consider an arbitrary function f of two variables, x and y. We apply the chain rule to obtain the derivative of f along a given curve with arc length s:

$$\frac{df(x(s),y(s))}{ds} = \frac{\partial f(x,y)}{\partial x}\frac{dx}{ds} + \frac{\partial f(x,y)}{\partial y}\frac{dy}{ds}. \tag{2.14}$$

This derivative is the directional derivative. The derivatives of f with respect to x and y are partial derivatives; f is a function of x and y; the derivatives with respect to s are ordinary derivatives because the curve along which we differentiate is defined by:

$$x = x(s) \quad y = y(s), \tag{2.15}$$

and thus x and y are functions of s only. We emphasize this in (2.14) by using parentheses to show the dependency of the function on its independent variables. We view f as a function of s only for the ordinary derivative to the left of the equal sign in (2.14), whereas we view it as a function of both x and y for the partial derivatives to the right of the equal sign. We omit this detail from here on, because it is sufficiently clear from the use of d and ∂ as differential operators, provided that we are precise in the choice of the differential operator. We are free to choose the direction of differentiation along any curve, which is fully defined by the expressions for dx/ds and dy/ds.

Since s is the length of arc along the curve, dx/ds and dy/ds are the components of the tangent unit vector along the curve (compare T_i in (2.5)). We generalize to three dimensions,

$$\frac{df}{ds} = \frac{\partial f}{\partial x}\frac{dx}{ds} + \frac{\partial f}{\partial y}\frac{dy}{ds} + \frac{\partial f}{\partial z}\frac{dz}{ds}, \tag{2.16}$$

and use indicial notation:

$$\frac{df}{ds} = \frac{\partial f}{\partial x_1}\frac{dx_1}{ds} + \frac{\partial f}{\partial x_2}\frac{dx_2}{ds} + \frac{\partial f}{\partial x_3}\frac{dx_3}{ds}. \tag{2.17}$$

We simplify this using the Einstein summation convention:

$$\frac{df}{ds} = \frac{\partial f}{\partial x_i}\frac{dx_i}{ds}. \tag{2.18}$$

We make a special choice for s; we no longer choose the arc length for s, but its projection onto one of the coordinate directions, for example the x_1 direction. We obtain for the case of (2.14):

$$\frac{df}{dx} = \frac{\partial f}{\partial x}\frac{dx}{dx} + \frac{\partial f}{\partial y}\frac{dy}{dx} = \frac{\partial f}{\partial x} + \frac{\partial f}{\partial y}\frac{dy}{dx}. \tag{2.19}$$

It is very important to recognize that df/dx is not the partial derivative of f with respect to x, but the derivative of f along s, multiplied by ds/dx. This is illustrated in Figure 2.3. We may obtain (2.19) by multiplying both sides of (2.14) by $ds/dx = 1/\cos\alpha$, where α is the inclination of the tangent of the curve to the x-axis. The direction of differentiation is now not represented via the values of the cosine and the sine of the angle α between the tangent to the curve and the x-axis ($\cos\alpha = dx/ds, \sin\alpha = dy/ds$), but by the tangent of the angle ($dy/dx = \tan\alpha$).

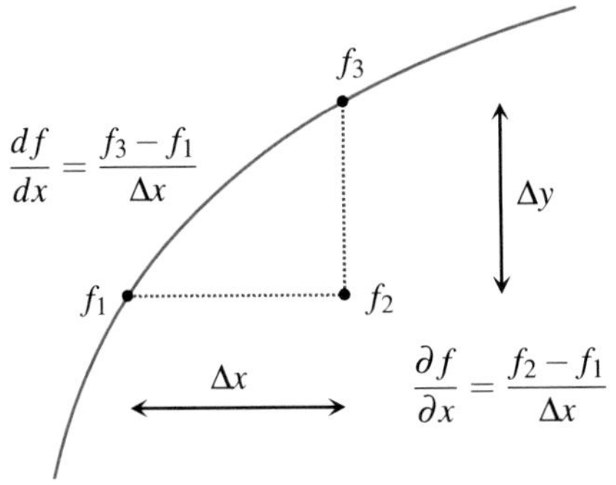

Figure 2.3: The directional derivative projected onto the x-direction

We return to the problem of determining the acceleration in the Eulerian description of the velocity field. The velocity is a function of space and time, and we may choose any curve in (x_k, t)-space (a four dimensional space). We introduce a parameter ξ such that the path is defined by $x_i = x_i(\xi)$ and $t = t(\xi)$, and apply the rule (2.18), with t added as an additional, but separate, dimension:

$$\frac{df}{d\xi} = \frac{\partial f}{\partial t}\frac{dt}{d\xi} + \frac{\partial f}{\partial x_i}\frac{dx_i}{d\xi}, \tag{2.20}$$

where the functions of the time and the coordinates in terms of the parameter ξ determine the direction of differentiation. We choose for ξ time, as we did choose x for s in (2.19):

$$\frac{df}{dt} = \frac{\partial f}{\partial t}\frac{dt}{dt} + \frac{\partial f}{\partial x_i}\frac{dx_i}{dt}. \tag{2.21}$$

We apply (2.20) to the velocity vector v_i in stead of an arbitrary function f:

$$\frac{dv_i}{dt} = \frac{\partial v_i}{\partial t} + \frac{\partial v_i}{\partial x_j}\frac{dx_j}{dt}. \tag{2.22}$$

We let the direction of differentiation coincide with the path of the fluid particle by setting dx_j/dt equal to the velocity v_j,

$$\frac{Dv_i}{Dt} = \frac{\partial v_i}{\partial t} + v_j \frac{\partial v_i}{\partial x_j}. \tag{2.23}$$

This derivative is a special derivative which is often used in physics and engineering and is called the material-time derivative; it represents the rate of change with time, while traveling through space with a particle, or material point; it is usually written with a capital D, rather than with d. The first term to the right of the equal sign in (2.23) represents the rate of change of the velocity vector with time, while being fixed in space, whereas the second term represents the rate of change of the velocity moving along the path, but with time fixed; the latter term is called the advective term.

For reasons that appear below, we modify the second term:

$$\frac{Dv_i}{Dt} = \frac{\partial v_i}{\partial t} + v_j \left[\frac{\partial v_i}{\partial x_j} - \frac{\partial v_j}{\partial x_i} \right] + v_j \frac{\partial v_j}{\partial x_i}, \tag{2.24}$$

or

$$\frac{Dv_i}{Dt} = \frac{\partial v_i}{\partial t} - v_j [\delta_{ip}\delta_{jq} - \delta_{jp}\delta_{iq}] \frac{\partial v_q}{\partial x_p} + v_j \frac{\partial v_j}{\partial x_i}. \tag{2.25}$$

We observe that

$$\frac{\partial (v_j v_j)}{\partial x_i} = \frac{\partial v^2}{\partial x_i} = v_j \frac{\partial v_j}{\partial x_i} + v_j \frac{\partial v_j}{\partial x_i} = 2 v_j \frac{\partial v_j}{\partial x_i}. \tag{2.26}$$

We use the ε - δ rule, (1.28), to rewrite (2.25), with $v_j v_j = v^2$,

$$\frac{Dv_i}{Dt} = a_i = \frac{\partial v_i}{\partial t} + \tfrac{1}{2} \frac{\partial v^2}{\partial x_i} - v_j \varepsilon_{ijk} \varepsilon_{kpq} \frac{\partial v_q}{\partial x_p}. \tag{2.27}$$

This equation represents the rate of change of the velocity along a path line, and thus represents the acceleration a_i. We use it for the derivation of one of the most important equations in applied fluid mechanics, the Bernoulli equation.

It can be shown that equations (2.27) and (2.10) are equivalent. The last term in (2.27) represents a component of the acceleration normal to the path line; the epsilon tensor represents the cross product; $\varepsilon_{ijk} v_j$ is perpendicular to v_j, i.e., perpendicular to the tangent to the path line.

We simplify our notation by writing ∂_i in stead of $\partial / \partial x_i$,

$$\partial_i = \frac{\partial}{\partial x_i}, \tag{2.28}$$

and rewrite (2.27):

$$a_i = \frac{\partial v_i}{\partial t} + \tfrac{1}{2} \partial_i v^2 - v_j \varepsilon_{ijk} \varepsilon_{kpq} \partial_p v_q. \tag{2.29}$$

If the flow is steady, the first term to the right of the equal sign vanishes, and the expression for the acceleration becomes

$$a_i = \tfrac{1}{2} \partial_i v^2 - \varepsilon_{ijk} v_j \varepsilon_{kpq} \partial_p v_q. \tag{2.30}$$

The path lines are curves that describe the path that the particles follow; each path line is fully identified by the initial condition of the particle, i.e., its position at a given time. The field lines of the vector field $v_i = v_i(x_k, t)$, are the stream lines; these lines are determined from the condition that the stream lines are tangent to the velocity vector, at each point at any time. The stream lines represent the instantaneous velocity field (that is why they are called field lines), whereas the path lines represent the path that the individual particles follow.

We are now ready to consider applications of Newton's second law; force equals mass times acceleration. We express the body force that acts on the fluid (gravity, for example) in mathematical terms, and then use expression (2.29) for the acceleration as it appears in this law.

Problem 2.1

Prove that the following equations hold

$$\vec{\nabla} \times \vec{\nabla}\phi = 0,$$

and

$$\vec{\nabla} \cdot \vec{\nabla} \times \vec{v} = 0,$$

where

$$\vec{\nabla} \equiv \partial_i \qquad (i = 1, 2, 3).$$

Problem 2.2

Under what condition(s) is the function $F = fg$ harmonic if both f and g are harmonic? Note: an harmonic function is a function that satisfies Laplace's equation, i.e, f is harmonic if $\partial_i \partial_i f = 0$.

Problem 2.3

Demonstrate that the function f is harmonic if

$$f = e_m \partial_m g,$$

where e_m is a fixed unit vector and the function g is harmonic.

Problem 2.4

Consider the function f, given by

$$f = a_m x_m b_n x_n,$$

where a_m and b_m are constants. What conditions must a_m and b_m satisfy for f to be harmonic?

Problem 2.5

Show that the function

$$f = \frac{e_i(x_i - \overset{0}{x}_i)}{[(x_m - \overset{0}{x}_m)(x_m - \overset{0}{x}_m)]^{3/2}}$$

is harmonic, where e_i represents a constant unit vector, and $\overset{0}{x}_i$ is a constant position vector.

Problem 2.6

Consider a fluid particle traveling with a constant velocity v (constant both in time and space) along a circle of radius R centered at the origin of a Cartesian coordinate system x_i. The circle (and thus the motion) is in the x_1,x_2 plane. Apply both the Lagrangian and Eulerian description of a moving particle to this case, and demonstrate that both reduce to the expression for centripetal acceleration and that the acceleration is in the direction of the origin.

Problem 2.7

A velocity field v_i is characterized by a magnitude of $v = 2$ m/s, an angle between the velocity and the horizontal plane $\alpha = 30^0$, and an angle between the horizontal projection of v_i and the x_1-axis equal to $\beta = 10^0$. Determine the discharge, in m^3/s, that flows through a triangular gate with corner points at $\overset{1}{x}_i = (0,5,4)$, $\overset{2}{x}_i = (3,0,8)$, and $\overset{3}{x}_i = (7,8,9)$.

Chapter 3

Fundamental Equations for Fluid Mechanics

3.1 The Euler and Bernoulli Equations

We consider force equilibrium of an elementary volume of fluid of constant density, $d\mathscr{V} = dx_1 dx_2 dx_3$, centered at x_i. The forces acting on $d\mathscr{V}$ are the forces on the six sides of the volume transmitted by the material outside $d\mathscr{V}$ and body forces, gravity, for example. We consider initially only cases where shear stresses transmitted via the viscosity of the fluid can be neglected; the stresses are isotropic, and reduce to a single scalar quantity[1], the pressure p. A resultant force in the x_1-direction exists if there is a difference between pressure acting on the face parallel to the (x_2, x_3)-plane centered at $x_1 - dx_1/2$ and that on the face centered at $x_1 + dx_1/2$. If we denote the pressures at the centers of these planes as $p(x_1 - dx_1/2, x_2, x_3)$ and $p(x_1 + dx_1/2, x_2, x_3)$, see Figure 3.1, then the component of the resultant force in the x_1-direction is $F_1 d\mathscr{V}$, with

$$F_1 d\mathscr{V} = -[p(x_1 + dx_1/2, x_2, x_3) - p(x_1 - dx_1/2, x_2, x_3)]dx_2 dx_3, \tag{3.1}$$

where F_1 is the resultant force per unit volume in the 1-direction. We divide both sides of this equation by $dx_1 dx_2 dx_3$,

$$F_1 = -\frac{[p(x_1 + dx_1/2, x_2, x_3) - p(x_1 - dx_1/2, x_2, x_3)]}{dx_1}, \tag{3.2}$$

and take the limit for $dx_1 \to 0$:

$$F_1 = -\frac{\partial p}{\partial x_1}. \tag{3.3}$$

We apply this process to all faces of the elementary volume to express the resultant force due to the pressure gradient at a point,

$$F_i = -\frac{\partial p}{\partial x_i}. \tag{3.4}$$

[1] A scalar is a quantity that has no orientation, as opposed to a vector.

O. D. L. Strack, *Applications of Vector Analysis and Complex Variables in Engineering*, https://doi.org/10.1007/978-3-030-41168-8_3

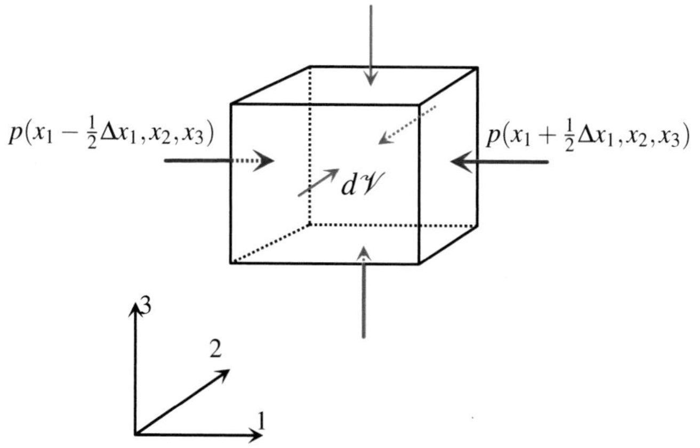

Figure 3.1: Forces acting on an elementary volume $d\mathscr{V}$

We consider the body force due to gravity, G_i,

$$G_i = -\rho g \delta_{3i}, \tag{3.5}$$

where ρ is the density and g the acceleration due to gravity. We apply Newton's second law:

$$\rho a_i = F_i + G_i, \tag{3.6}$$

and use (3.4) and (3.5)

$$\rho a_i = -\partial_i p - \rho g \partial_i x_3. \tag{3.7}$$

We used that $\partial x_3 / \partial x_i$ is non-zero only if $i = 3$, or, for the general case

$$\delta_{ij} = \frac{\partial x_j}{\partial x_i} = \partial_i x_j. \tag{3.8}$$

We use (2.29) to express a_i in terms of the velocity components, combine this with (3.7) and divide by ρg:

$$\frac{1}{g}\frac{\partial v_i}{\partial t} + \frac{1}{2g}\frac{\partial v^2}{\partial x_i} - \frac{1}{g}\varepsilon_{ijk} v_j \varepsilon_{kpq} \frac{\partial v_q}{\partial x_p} = -\frac{1}{\rho g}\frac{\partial p}{\partial x_i} - \frac{\partial x_3}{\partial x_i}. \tag{3.9}$$

This is Euler's equation, obtained by application of Newton's second law to an infinitesimal fluid body.

We project Euler's equation onto the direction of flow by forming the dot product of the vectors in (3.9) and the tangent vector to the streamline, T_i,[2] and restrict ρ to be constant,

$$\frac{1}{g}\frac{\partial v_i}{\partial t}T_i + \frac{1}{2g}\frac{\partial v^2}{\partial x_i}T_i - \frac{1}{g}\varepsilon_{ijk} v_j \varepsilon_{kpq} \frac{\partial v_q}{\partial x_p}T_i = -\frac{\partial\left[\frac{p}{\rho g} + x_3\right]}{\partial x_i}T_i. \tag{3.10}$$

[2]This process is called contraction with the unit vector T_i

Since $\varepsilon_{ijk}v_j$ is perpendicular to v_i, and T_i is tangent to v_i, the dot product of these two quantities is zero for all k. We use (2.5), to write T_i as dx_i/ds, in (3.10):

$$\frac{1}{g}\frac{\partial v_i}{\partial t}T_i + \frac{1}{2g}\frac{\partial v^2}{\partial x_i}\frac{dx_i}{ds} = -\frac{\partial\left[\frac{p}{\rho g} + x_3\right]}{\partial x_i}\frac{dx_i}{ds}, \tag{3.11}$$

or

$$\frac{1}{g}\frac{\partial v_i}{\partial t}T_i + \frac{1}{2g}\frac{dv^2}{ds} = -\frac{d\left[\frac{p}{\rho g} + x_3\right]}{ds}. \tag{3.12}$$

The first term vanishes for steady flow. The following equation is valid for constant density, steady flow, and along a streamline,

$$\frac{1}{2g}\frac{dv^2}{ds} + \frac{d(\frac{p}{\rho g} + x_3)}{ds} = 0. \tag{3.13}$$

We integrate this equation along a streamline, and obtain Bernoulli's equation:

$$\frac{v^2}{2g} + \frac{p}{\rho g} + x_3 = H. \tag{3.14}$$

The quantity H is the energy head; it is constant along streamlines, but generally varies in other directions. The energy head H [L] is the energy divided by ρg, and:

$$(p + \rho g x_3) + \tfrac{1}{2}\rho v^2 = \rho g H. \tag{3.15}$$

The term in parentheses represents the potential energy per unit volume and the second term is the kinematic energy per unit volume; $\rho g H$ is the energy stored in the fluid (with the exception of internal energy) per unit volume. Equation (3.14) in words reads: velocity head plus pressure head and elevation head equals energy head.

It is not surprising that the energy head is constant along streamlines; viscosity is neglected, and there are no energy losses due to shear stresses in the fluid. It is surprising, however, that energy can be dissipated even if the viscosity is neglected. This dissipation occurs in directions other than tangent to the streamlines if the third term to the left of the equal sign in (3.9) does not vanish. We will see that this term is non-zero only if the flow is rotational.

An important, but elementary, case applies when the flow is uniform and horizontal. In that case, there is no component of acceleration in the direction normal to the flow and therefore the force balance equation (3.7) in the vertical direction reduces to

$$\frac{\partial\left[\frac{p}{\rho g} + x_3\right]}{\partial x_3} = 0, \tag{3.16}$$

so that the hydraulic head ϕ is a constant in the vertical direction

$$\phi = \frac{p}{\rho g} + x_3 = C, \tag{3.17}$$

where C is a constant.

3.2 The Mass Balance Equation

An important equation that must be satisfied is the mass balance equation, which involves a scalar quantity, the divergence. The divergence of a vector v_i is:

$$\frac{\partial v_i}{\partial x_i} = \partial_i v_i \equiv \frac{\partial v_x}{\partial x} + \frac{\partial v_y}{\partial y} + \frac{\partial v_z}{\partial z} = D. \tag{3.18}$$

It is called the divergence because a positive value of D will, in the absence of other effects, cause a flow that diverges from a point. To understand the meaning of the divergence, consider the case of flow in one direction only, the x_1-direction; the divergence D reduces to:

$$D = \frac{dv_x}{dx}. \tag{3.19}$$

We consider an increment dx, see Figure 3.2, and obtain as an approximation of $Dd\mathcal{V}$, where $d\mathcal{V} = dx_1 dx_2 dx_3$,

$$Dd\mathcal{V} = Ddxdydz \approx [v_x(x+dx/2) - v_x(x-dx/2)]dydz. \tag{3.20}$$

This represents the amount of fluid that leaves the elementary volume $d\mathcal{V}$ per unit time, and must be zero, unless either fluid is added, or the volume is compressed. In

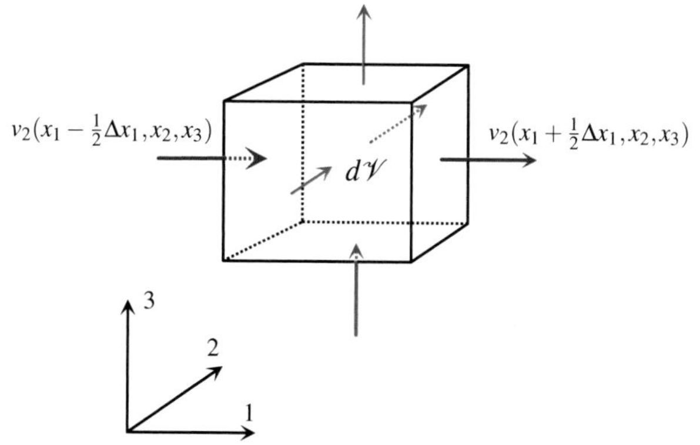

Figure 3.2: The divergence of the velocity vector

the case of three-dimensional flow, a non-zero divergence indicates that an amount D of fluid is produced per unit volume and per unit time. If the fluid is incompressible, the divergence of the velocity vector is zero:

$$\frac{\partial v_i}{\partial x_i} = 0. \tag{3.21}$$

If the fluid is compressible, the divergence is the amount of mass produced per unit time and per unit volume. The divergence of ρv_i is equal to the rate of decrease of mass at a point,

$$\frac{\partial \rho v_i}{\partial x_i} = -\frac{\partial \rho}{\partial t}. \tag{3.22}$$

Note that the minus sign indicates that a positive divergence corresponds to a decrease in mass, i.e., a negative value of the rate of change of the density with time.

Most liquids can with good approximation be considered incompressible.

3.3 Rotational Flow

We consider an elementary volume centered at x_i and with sides dx_i and volume $d\mathscr{V} = dx_1 dx_2 dx_3$. Rotation occurs if the tangential components of flow give a rotational resultant motion. We represent the magnitude of rotation per unit volume and per unit time as $\omega = C$, see Figure 3.3, where ω is the angular rotation of the fluid. Consider the face with normal in the x_3 direction, and sides dx_1 and dx_2. The rotation in the x_1, x_2 plane is equal to

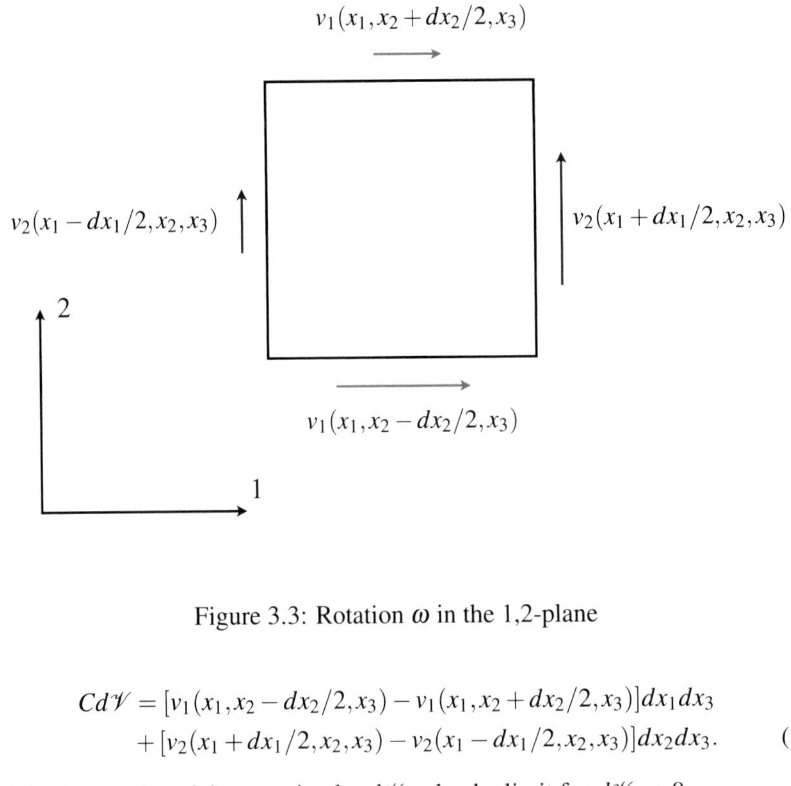

Figure 3.3: Rotation ω in the 1,2-plane

$$\begin{aligned} C d\mathscr{V} &= [v_1(x_1, x_2 - dx_2/2, x_3) - v_1(x_1, x_2 + dx_2/2, x_3)] dx_1 dx_3 \\ &+ [v_2(x_1 + dx_1/2, x_2, x_3) - v_2(x_1 - dx_1/2, x_2, x_3)] dx_2 dx_3. \end{aligned} \tag{3.23}$$

We divide both sides of the equation by $d\mathscr{V}$, take the limit for $d\mathscr{V} \to 0$:

$$C_3 = -\frac{\partial v_1}{\partial x_2} + \frac{\partial v_2}{\partial x_1}. \tag{3.24}$$

We represent the rotation by a vector; the component that represents counterclockwise rotational flow in the x_1, x_2-plane points vertically upward (the corkscrew rule), see

Figure 3.4. We repeat the procedure for computing the rotational components in the two remaining planes and obtain

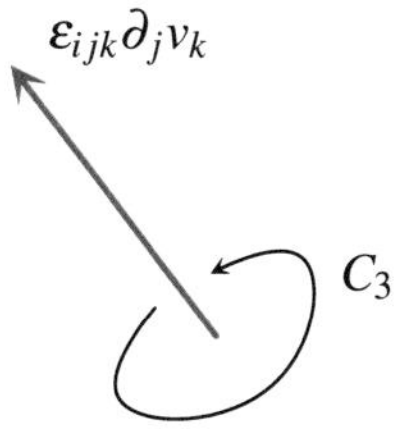

Figure 3.4: The curl as a vector

$$C_i = \varepsilon_{ijk}\partial_j v_k. \tag{3.25}$$

The expression on the right is called the curl of the velocity vector. We verify that this expression for C_i reduces to (3.24), for $i = 3$,

$$C_3 = \varepsilon_{312}\partial_1 v_2 + \varepsilon_{321}\partial_2 v_1 = \frac{\partial v_2}{\partial x_1} - \frac{\partial v_1}{\partial x_2}, \tag{3.26}$$

which indeed matches (3.24).

3.4 The Bernoulli Equation for Irrotational Flow

The flow is irrotational if the curl of the velocity vector vanishes, i.e., if

$$C_i = \varepsilon_{ijk}\partial_j v_k = 0. \tag{3.27}$$

The last term to the left of the equal sign in Euler's equation (3.9) vanishes for irrotational flow,

$$\frac{1}{g}\frac{\partial v_i}{\partial t} + \frac{1}{2g}\frac{\partial v^2}{\partial x_i} = -\frac{1}{\rho g}\frac{\partial p}{\partial x_i} - \frac{\partial x_3}{\partial x_i}. \tag{3.28}$$

If the flow is steady and ρ is constant, this becomes

$$\frac{\partial}{\partial x_i}\left[\frac{v^2}{2g} + \frac{p}{\rho g} + x_3\right] = 0. \tag{3.29}$$

This equation is valid throughout the field; integration is possible, and we obtain Bernoulli's equation for irrotational flow:

$$\frac{v^2}{2g} + \frac{p}{\rho g} + x_3 = H \tag{3.30}$$

where H is constant throughout the flow domain, as opposed to (3.14), which is valid only along streamlines.

3.4.1 Conservation of Zero Rotation

Equation (3.28) is valid for irrotational transient flow. We examine the rotation of a velocity field that is, at some time t_0, irrotational. Such a velocity field satisfies (3.28) at $t = t_0$,

$$\frac{1}{g}\frac{\partial v_i}{\partial t} + \frac{1}{2g}\frac{\partial v^2}{\partial x_i} + \frac{1}{\rho g}\frac{\partial p}{\partial x_i} + \frac{\partial x_3}{\partial x_i} = A_i = 0. \tag{3.31}$$

We replace i by k and take the curl:

$$\frac{1}{g}\varepsilon_{ijk}\partial_j\frac{\partial v_k}{\partial t} + \frac{1}{2g}\varepsilon_{ijk}\partial_j\partial_k v^2 + \varepsilon_{ijk}\partial j\partial_k\left[\frac{p}{\rho g} + x_3\right] = 0. \tag{3.32}$$

The operator $\partial_j\partial_k$ is symmetrical; the order of differentiation is immaterial, and the alternating tensor vanishes if operated on a term that is symmetrical in two of its indices. Thus, the second and third terms in (3.32) vanish. The order of differentiation with respect to t and x_j in the first term is nearly always interchangeable and under such conditions:

$$\frac{\partial \varepsilon_{ijk}\partial_j v_k}{\partial t} = 0, \tag{3.33}$$

or

$$\varepsilon_{ijk}\partial_j v_k = \underset{r}{C_i}, \tag{3.34}$$

where the components of the curl $\underset{r}{C_i}$ are constants. This shows that if the rotation is initially zero, then it will remain zero. We cannot conclude from this result that if the rotation is constant at time t_0, it will remain constant; (3.34) is valid only if the rotation is initially zero.

3.4.2 Transient Flow of a Compressible Inviscid Fluid with Rotation

Euler's equation (3.9) is valid for general flow of an inviscid fluid:

$$\frac{1}{g}\frac{\partial v_i}{\partial t} + \frac{1}{2g}\frac{\partial v^2}{\partial x_i} - \frac{1}{g}\varepsilon_{ijk}v_j\varepsilon_{kpq}\frac{\partial v_q}{\partial x_p} = -\frac{1}{\rho g}\frac{\partial p}{\partial x_i} - \frac{\partial x_3}{\partial x_i}. \tag{3.35}$$

We consider a compressible fluid, with the density depending only on pressure. We accommodate such variable density by introducing the following function P:

$$P = \int \frac{1}{\rho(p)g}dp, \tag{3.36}$$

so that

$$\partial_i P = \frac{dP}{dp}\frac{\partial p}{\partial x_i} = \frac{1}{\rho g}\partial_i p. \tag{3.37}$$

We use this to rewrite (3.35):

$$\frac{\partial}{\partial x_i}\left[P + x_3 + \frac{v^2}{2g}\right] = -\frac{1}{g}\frac{\partial v_i}{\partial t} + \frac{1}{g}\varepsilon_{ijk}v_j\varepsilon_{kpq}\partial_p v_q. \tag{3.38}$$

We write $\varepsilon_{kpq}\partial_p v_q$ as C_k, which is the curl,

$$\frac{\partial}{\partial x_i}\left[P+x_3+\frac{v^2}{2g}\right] = -\frac{1}{g}\frac{\partial v_i}{\partial t}+\frac{1}{g}\varepsilon_{ijk}v_jC_k, \tag{3.39}$$

and take the divergence of this expression,

$$\frac{\partial^2}{\partial x_i \partial x_i}\left[P+x_3+\frac{v^2}{2g}\right] = -\frac{1}{g}\frac{\partial[\partial_i v_i]}{\partial t}+\frac{1}{g}\varepsilon_{ijk}\partial_i[v_jC_k]. \tag{3.40}$$

We differentiate the last term by parts and rearrange indices,

$$\frac{1}{g}\varepsilon_{ijk}\partial_i[v_jC_k] = \frac{1}{g}\varepsilon_{ijk}\partial_i v_j C_k + \frac{1}{g}\varepsilon_{ijk}v_j\partial_i C_k = \frac{1}{g}\varepsilon_{kij}\partial_i v_j C_k - \frac{1}{g}v_j\varepsilon_{jik}\partial_i C_k. \tag{3.41}$$

The first term to the right of the equal sign is equal to $C_kC_k = C^2$. We often refer to the magnitude of the curl, C, as the rotation, ω,

$$\omega = C = \sqrt{C_kC_k}. \tag{3.42}$$

The last term in (3.41) is the dot product of the velocity vector and the curl of the curl of the velocity vector; (3.40) becomes, after renaming indices for clarity,

$$\frac{\partial^2}{\partial x_i \partial x_i}\left[P+x_3+\frac{v^2}{2g}\right] = -\frac{1}{g}\frac{\partial[\partial_i v_i]}{\partial t}+\frac{C^2}{g}-\frac{1}{g}v_i\varepsilon_{ijk}\partial_j C_k. \tag{3.43}$$

The term between the brackets to the left of the equal sign is the energy head H, and $\partial^2/(\partial x_i \partial x_i) = \partial_i\partial_i$ is the Laplacian ∇^2 so that

$$\nabla^2 H = -\frac{1}{g}\frac{\partial[\partial_i v_i]}{\partial t}+\frac{C^2}{g}-\frac{1}{g}v_i\varepsilon_{ijk}\partial_j C_k, \tag{3.44}$$

where

$$H = P+x_3+\frac{v^2}{2g}. \tag{3.45}$$

We write the last term in (3.44) in a different manner and obtain as an alternative expression

$$\nabla^2 H = -\frac{1}{g}\frac{\partial[\partial_i v_i]}{\partial t}+\frac{C^2}{g}-\frac{1}{g}v_i\varepsilon_{ijk}\varepsilon_{kpq}\partial_j\partial_p v_q, \tag{3.46}$$

or,

$$\nabla^2 H = -\frac{1}{g}\frac{\partial[\partial_i v_i]}{\partial t}+\frac{C^2}{g}-v_i[\delta_{ip}\delta_{jq}-\delta_{iq}\delta_{jp}]\partial_j\partial_p v_q. \tag{3.47}$$

We expand the last two terms and obtain

$$\nabla^2 H = -\frac{1}{g}\frac{\partial[\partial_i v_i]}{\partial t}+\frac{C^2}{g}-v_i\partial_i\partial_j v_j+v_i\partial_j\partial_j v_i. \tag{3.48}$$

If the flow is divergence-free, then the terms that contain the divergence $\partial_j v_j$ vanish, and

$$\nabla^2 H = \frac{C^2}{g} + v_i \nabla^2 v_i = \frac{C^2}{g} - \frac{1}{g} v_i \varepsilon_{ijk} \partial_j C_k. \tag{3.49}$$

A function with a Laplacian unequal to zero cannot be equal to a constant throughout the domain; if the rotation ω is unequal to zero, the energy head varies throughout the field and the streamlines are the contours of constant H. For transient incompressible flow of an inviscid fluid, rotation is the only way for energy to change in the system. We see in Chapter 13 that the same is true even if the viscosity cannot be neglected.

This relation between energy head and rotation is typical for fluid mechanics; in other fields, such as conduction of heat in solids or groundwater flow, this relation does not hold, see Chapter 10. In the case of groundwater flow, for example, Euler's equation applies, but the acceleration is negligible; energy loss occurs due to friction between water and solid particles, rather than due to rotation.

3.5 Objects Moving through the Fluid

A special case of transient flow is that of an object moving through the fluid, with the fluid far away at rest. Such problems can be solved by attaching the coordinate system to the moving object, defining the velocities relative to the moving coordinate system, as shown in Figure 3.5. We write Euler's equation in terms of the moving coordinate

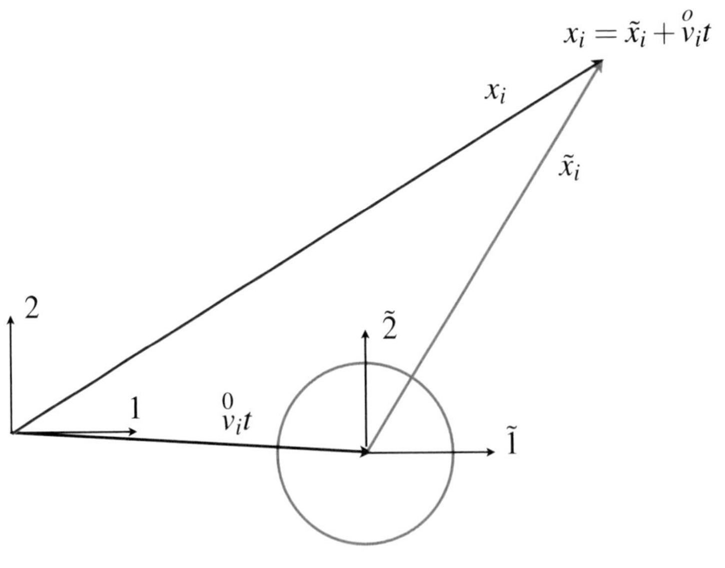

Figure 3.5: Relation between local and global coordinates

system; we express the acceleration of an elementary fluid volume in terms of the moving coordinates. We consider a moving sphere as an example; the sphere moves with a constant velocity $\overset{0}{v}_i$. Even if the fluid were at rest, it would move with a velocity $-\overset{0}{v}_i$ as viewed from the moving coordinate system. We label the moving coordinates as $\tilde{x}_i$, and the coordinate transformation is

$$\tilde{x}_i = x_i - \overset{0}{v}_i(t - t_0), \tag{3.50}$$

so that the coordinate systems coincide at $t = t_0$. The center of the sphere is at $\tilde{x}_i = 0$, so that $x_i = \overset{0}{v}_i(t - t_0)$ represents the coordinates of the center of the sphere. If the center of the object is at the origin of the coordinate system at time $t = 0$, then the relation between local and global coordinates is, see Figure 3.5,

$$x_i = \tilde{x}_i + \overset{o}{v}_i t. \tag{3.51}$$

We write the velocity vector with respect to the moving coordinates as $\tilde{v}_i$ and consider the material-time derivative of $v_i = \tilde{v}_i + \overset{0}{v}_i$:

$$\frac{Dv_i}{Dt} = \frac{\partial \tilde{v}_i}{\partial t} + \frac{\partial \overset{o}{v}_i}{\partial t} + \frac{\partial(\tilde{v}_i + \overset{o}{v}_i)}{\partial x_j}\frac{dx_j}{dt}. \tag{3.52}$$

The velocity of the sphere, $\overset{o}{v}_i$, is constant; $\partial \overset{o}{v}_i/\partial t = 0$ and $\partial \overset{o}{v}_i/\partial x_j = 0$,

$$\frac{Dv_i}{Dt} = \frac{\partial \tilde{v}_i}{\partial t} + \frac{\partial \tilde{v}_i}{\partial x_j}\frac{dx_j}{dt}. \tag{3.53}$$

We differentiate (3.50) with respect to time:

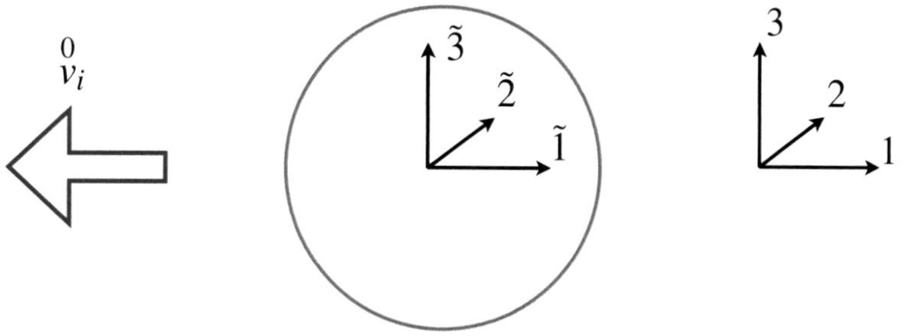

Figure 3.6: An object moving through the fluid

$$\frac{d\tilde{x}_i}{dt} = \frac{dx_i}{dt} - \overset{o}{v}_i \rightarrow \frac{dx_i}{dt} = \tilde{v}_i + \overset{0}{v}_i, \tag{3.54}$$

so that (3.53) becomes:

$$\frac{Dv_i}{Dt} = \frac{\partial \tilde{v}_i}{\partial t} + \frac{\partial \tilde{v}_i}{\partial x_j}(\tilde{v}_j + \overset{o}{v}_j), \tag{3.55}$$

or

$$\frac{Dv_i}{Dt} = \frac{\partial \tilde{v}_i}{\partial t} + \frac{\partial \tilde{v}_i}{\partial x_j} \overset{o}{v}_j + \frac{\partial \tilde{v}_i}{\partial x_j} \tilde{v}_j. \tag{3.56}$$

The first two terms to the right of the equal sign together represent the rate of change with time of the relative velocity $\tilde{v}_i$ in the moving coordinate system, i.e., viewed from a point that moves with velocity $\overset{o}{v}_i$,

$$\frac{Dv_i}{Dt} = \frac{d\tilde{v}_i}{dt} + \tilde{v}_j \frac{\partial \tilde{v}_i}{\partial x_j}. \tag{3.57}$$

The first term to the right of the equal sign is to be evaluated as if it were a partial derivative, i.e., by letting the coordinates $\tilde{x}_i$ be constant in the expression for $\tilde{v}_i$. The derivative of $\tilde{v}_i$ with respect to x_j can be replaced with the derivative with respect to $\tilde{x}_j$, because the two coordinate systems differ only by a function of time, i.e.,

$$\frac{Dv_i}{Dt} = \frac{d\tilde{v}_i}{dt} + \tilde{v}_j \frac{\partial \tilde{v}_i}{\partial \tilde{x}_j}. \tag{3.58}$$

3.5.1 Euler's Equation for Moving Coordinates; Irrotational Flow

We limit the analysis to irrotational flow, and write Euler's equation in terms of the moving coordinates. The flow appears to be steady as viewed from the origin of the moving coordinate system. We apply Newton's second law to expression (3.58) for the acceleration

$$\rho \left[\frac{d\tilde{v}_i}{dt} + \tilde{v}_j \frac{\partial \tilde{v}_i}{\partial \tilde{x}_j} \right] = -\frac{\partial}{\partial \tilde{x}_i} \left[p + \rho g x_3 \right], \tag{3.59}$$

and rewrite the second term inside the brackets as we did in equations (2.24) through (2.27) on page 15,

$$\tilde{v}_j \frac{\partial \tilde{v}_i}{\partial \tilde{x}_j} = \tfrac{1}{2} \frac{\partial \tilde{v}^2}{\partial \tilde{x}_i} - \tilde{v}_j \varepsilon_{ijk} \tilde{C}_k, \tag{3.60}$$

where $\tilde{C}_k = C_k$ is the curl ($C_k = \tilde{C}_k$, because $\overset{o}{v}_i$ is constant). In the present case of irrotational flow, this reduces to

$$\tilde{v}_j \frac{\partial \tilde{v}_i}{\partial \tilde{x}_j} = \tfrac{1}{2} \frac{\partial \tilde{v}^2}{\partial \tilde{x}_i}. \tag{3.61}$$

We apply this to (3.59) and obtain, upon division by ρg:

$$\frac{1}{g} \frac{d\tilde{v}_i}{dt} + \frac{\partial}{\partial \tilde{x}_i} \left[\frac{1}{2g} \tilde{v}^2 + \frac{p}{\rho g} + x_3 \right] = 0. \tag{3.62}$$

Since the object moves with constant velocity, $\tilde{v}_i$ does not change with time, viewed from the moving object, i.e.,

$$\frac{d\tilde{v}_i}{dt} = \frac{\partial \tilde{v}_i}{\partial t} + \frac{\partial \tilde{v}_i}{\partial \tilde{x}_j} \overset{0}{v}_j = 0. \tag{3.63}$$

The partial derivative of $\tilde{v}_i$ with respect to time is not zero, but equal to a function of position

$$\frac{\partial \tilde{v}_i}{\partial t} = -\frac{\partial \tilde{v}_i}{\partial \tilde{x}_j}\overset{0}{v}_j, \tag{3.64}$$

and the partial derivatives of v_i and $\tilde{v}_i$ with respect to time are equal because $\overset{0}{v}_i$ is constant,

$$\frac{\partial \tilde{v}_i}{\partial t} = \frac{\partial (v_i - \overset{0}{v}_i)}{\partial t} = \frac{\partial v_i}{\partial t}. \tag{3.65}$$

We use (3.63) in Euler's equation, (3.62),

$$\frac{\partial}{\partial \tilde{x}_i}\left[\frac{1}{2g}\tilde{v}^2 + \frac{p}{\rho g} + x_3\right] = 0. \tag{3.66}$$

The function inside the brackets is not the energy head, since it contains $\tilde{v}$ rather than the actual velocity. This function, $\tilde{H}$, does not depend on $\tilde{x}_i$ and thus is a constant both in space and time, as it is viewed from the origin of the moving coordinate system,

$$\tilde{H} = \frac{1}{2g}\tilde{v}^2 + \frac{p}{\rho g} + x_3 = C^*, \tag{3.67}$$

where C^* is a constant in $\tilde{x}_i$ space. Note that $\tilde{H}$ is not the energy head expressed in terms of the moving coordinates; the velocity $\tilde{v}_i$ is not equal to the velocity, so that $\tilde{H}$ differs from the actual energy head.

3.5.2 The Energy Head H

The energy head for this case is not a constant, because the velocity field depends on time; this becomes clear by comparing the function $\tilde{H}$, which is a constant, with the actual energy head, which is:

$$H = \frac{1}{2g}v^2 + \frac{p}{\rho g} + x_3. \tag{3.68}$$

The modulus, or magnitude, of the velocity vector $\tilde{v}_i$ is related to that of the velocity v_i as

$$\tilde{v}^2 = \tilde{v}_i\tilde{v}_i = (v_i - \overset{0}{v}_i)(v_i - \overset{0}{v}_i) = v_iv_i - 2\overset{0}{v}_iv_i + \overset{0}{v}_i\overset{0}{v}_i = v^2 - 2\overset{0}{v}_iv_i + \overset{0}{v}{}^2. \tag{3.69}$$

We use this equation to express $\tilde{H}$, (3.67), in terms of H:

$$\tilde{H} = \frac{1}{2g}\tilde{v}^2 + \frac{p}{\rho g} + x_3 = \frac{1}{2g}v^2 + \frac{p}{\rho g} + x_3 - \frac{1}{2g}\left[2\overset{0}{v}_iv_i - \overset{0}{v}{}^2\right] = C^*, \tag{3.70}$$

so that

$$H = \tilde{H} + \frac{1}{2g}\left[2\overset{0}{v}_iv_i - \overset{0}{v}{}^2\right]. \tag{3.71}$$

We present an alternative derivation of the latter expression for the energy head H using global coordinates for the sake of completeness, reproduced below in small print, as optional reading.

Derivation of the Expression for H Using Global Coordinates

We obtain the equation for the energy head that applies to this case in terms of global coordinates from (3.28) on page 23,

$$\frac{1}{g}\frac{\partial v_i}{\partial t}+\frac{1}{2g}\frac{\partial v^2}{\partial x_i}=-\frac{1}{\rho g}\frac{\partial p}{\partial x_i}-\frac{\partial x_3}{\partial x_i}, \tag{3.72}$$

or

$$\partial_i\left[\frac{v^2}{2g}+\frac{1}{\rho g}p+x_3\right]=-\frac{1}{g}\frac{\partial v_i}{\partial t}. \tag{3.73}$$

We use (3.64) and (3.65) to rewrite the term to the right of the equal sign and replace the partial derivatives with respect to $\tilde{x}_i$ by those with respect to x_i:

$$\partial_i\left[\frac{v^2}{2g}+\frac{p}{\rho g}+x_3\right]=\frac{1}{g}\overset{0}{v}_j\partial_j\tilde{v}_i. \tag{3.74}$$

We use the following relationship:

$$\begin{aligned}\partial_j\tilde{v}_i&=\partial_j\tilde{v}_i-\partial_i\tilde{v}_j+\partial_i\tilde{v}_j=-(\delta_{ip}\delta_{jq}-\delta_{jp}\delta_{iq})\partial_p\tilde{v}_q+\partial_i\tilde{v}_j\\&=-\varepsilon_{mij}\varepsilon_{mpq}\partial_p\tilde{v}_q+\partial_i\tilde{v}_j=-\varepsilon_{mij}\tilde{C}_m+\partial_i\tilde{v}_j,\end{aligned} \tag{3.75}$$

where $\tilde{C}_m$ is the curl of the velocity field. The velocity field is irrotational, so that the curl vanishes, and we obtain for (3.74)

$$\partial_i\left[\frac{v^2}{2g}+\frac{p}{\rho g}+x_3\right]=\frac{1}{g}\overset{0}{v}_j\partial_j\tilde{v}_i=\frac{1}{g}\overset{0}{v}_j\partial_i\tilde{v}_j=\frac{1}{g}\partial_i(\overset{0}{v}_j\tilde{v}_j), \tag{3.76}$$

so that

$$H=\frac{1}{g}\overset{0}{v}_j\tilde{v}_j+C, \tag{3.77}$$

where C is a constant. This equation corresponds to the result obtained by comparing the energy head H with $\tilde{H}$, (3.71).

We may describe the flow caused by an object that moves through a fluid at a constant speed by treating the object as stationary and the flow as steady, provided that the velocity is computed relative to the object. Remember that these velocity values are relative, and that the actual velocity distribution is obtained by adding the velocity of the object, $\overset{o}{v}_i$, to the velocity vector $\tilde{v}_i$.

The energy head is a function of position, given by

$$H=\tilde{H}+\frac{1}{g}v_j\overset{o}{v}_j-\frac{\overset{o}{v}{}^2}{2g}, \tag{3.78}$$

where $\tilde{H}$ is a constant. If the fluid is at rest at infinity, then v_j approaches zero there, so that

$$H\to\tilde{H}-\frac{\overset{o}{v}{}^2}{2g}\qquad x_ix_i\to\infty, \tag{3.79}$$

which is the lowest value for H possible for this case. The two expressions for H, (3.77) and (3.79), differ by the constant expression $\overset{o}{v}{}^2/(2g)$.

Problem 3.1

This exercise deals with a two-dimensional vector field; there is no flow in the x_3-direction, and the indices now run from 1 to 2. The Kronecker delta is defined in the same way as for the three-dimensional case, but the ε tensor is now defined with two indices: $\varepsilon_{ij} = 0$ if $i = j$ and $\varepsilon_{12} = -\varepsilon_{21} = 1$. Consider a two-dimensional velocity field v_i defined as

$$v_i = \overset{o}{v}\left[\delta_{1i}\left(1+\frac{R^2}{r^2}\right) - 2\frac{x_1 x_i R^2}{r^4}\right]. \tag{3.80}$$

This velocity field describes flow around an impermeable cylinder centered at $x_i = 0$ and with radius R. The flow far away from the cylinder has velocity $\overset{o}{v}$ and is in the x_1 direction. The distance from the origin to the point x_i is r, so that

$$r^2 = x_i x_i = x_1^2 + x_2^2 \equiv x^2 + y^2. \tag{3.81}$$

The axis of the cylinder is normal to the plane of flow.

Questions

1. Demonstrate that the velocity is tangent to the cylinder wall.
2. Demonstrate that the velocity is zero at $x_i = (R,0)$ and at $x_i = (-R,0)$.
3. Compute the tangential and normal acceleration at $x_i = (0,-R)$.

3.6 Flow with Divergence and Rotation; the General Field

If the velocity vector can be written as the gradient of a scalar velocity potential, i.e., if

$$v_i = -\partial_i \Phi, \tag{3.82}$$

then the resulting vector field is irrotational. We verify this by taking the curl of the velocity vector,

$$\varepsilon_{ijk}\partial_j v_k = \varepsilon_{ijk}\partial_j\partial_k\Phi, \tag{3.83}$$

which is zero because the differential operator $\partial_j\partial_k$ is symmetrical (the order of differentiation is immaterial), so that

$$\varepsilon_{ijk}\partial_j v_k = 0. \tag{3.84}$$

This proves that the gradient of a scalar potential is irrotational; it does not prove that any irrotational velocity field can be represented as the gradient of a scalar potential.

If we represent the velocity vector as the curl of a function W_k, called a vector potential, i.e.,

$$v_i = -\varepsilon_{ijk}\partial_j W_k, \tag{3.85}$$

then the divergence of this vector field is zero:

$$\partial_i v_i = \varepsilon_{ijk}\partial_i\partial_j W_k = 0. \tag{3.86}$$

If the velocity field has a divergence $-\gamma$ and rotation δ then we can attempt to represent v_i as:

$$v_i = -\partial_i\Phi - \varepsilon_{ijk}\partial_j W_k. \tag{3.87}$$

We take the divergence of this vector field, which gives:

$$\partial_i v_i = -\partial_i\partial_i\Phi = -\nabla^2\Phi = D = -\gamma \rightarrow \nabla^2\Phi = \gamma, \tag{3.88}$$

and take the curl of v_i:

$$\varepsilon_{ijk}\partial_j v_k = -\varepsilon_{ijk}\varepsilon_{kpq}\partial_j\partial_p W_q = -[\delta_{ip}\delta_{jq} - \delta_{iq}\delta_{jp}]\partial_j\partial_p W_q = -\partial_i\partial_p W_p + \nabla^2 W_i. \tag{3.89}$$

The first term to the right of the final equal sign is the gradient of the divergence D of the vector potential W_i; it is always possible to choose W_i in such a manner that it is itself divergence-free. To see this, consider the case that a function $\bar{W}_i$ is a solution to (3.85) with divergence $\partial_i\bar{W}_i = \beta$. We choose a second vector potential G_i such that it is irrotational, which we can do by choosing $G_i = \partial_i F$. We choose F such that $\partial_i G_i = \beta$ and subtract G_i from the function $\bar{W}_i$, so that

$$W_i = \bar{W}_i - G_i. \tag{3.90}$$

This function still satisfies (3.85) because $\varepsilon_{ijk}\partial_j G_k = \varepsilon_{ijk}\partial_j\partial_k F = 0$ so that the additional vector potential G_i does not contribute to the velocity vector. The divergences of $\bar{W}_i$ and G_i are equal; they cancel in the divergence of the vectors in (3.90), so that the vector potential W_i is indeed divergence free. We choose W_i such that

$$\partial_i W_i = 0, \tag{3.91}$$

and equation (3.89) becomes

$$\varepsilon_{ijk}\partial_j v_k = C_i = \nabla^2 W_i. \tag{3.92}$$

It can be shown that a divergence-free vector field can always be represented as the curl of a vector potential, and that an irrotational vector field can always be represented as the gradient of a scalar potential. The proof of this statement is non-trivial and will not be given here; the reader is referred to the literature for a formal proof.

An important theorem, known as Helmholtz's decomposition theorem, states that any general vector field can be decomposed into a divergence-free vector field represented as the curl of a vector potential, and an irrotational vector field, represented as the gradient of a scalar potential. Thus, any vector field v_i can be represented as

$$v_i = -\partial_i\Phi - \varepsilon_{ijk}\partial_j W_k - \partial_i F \tag{3.93}$$

where

$$\begin{aligned}\nabla^2\Phi &= -D\\ \nabla^2 W_i &= C_i\\ \nabla^2 F &= 0\end{aligned} \tag{3.94}$$

where D is the divergence and C_i the curl.

If the velocity potential exists, it can be used to integrate the expression for the energy head H for transient flow, see (3.38) on page 24. The latter equation becomes, using the energy head H and setting the curl equal to zero,

$$\partial_i H = -\frac{1}{g}\frac{\partial v_i}{\partial t} = \partial_i\left[\frac{1}{g}\frac{\partial\Phi}{\partial t}\right], \tag{3.95}$$

so that

$$H = \frac{1}{g}\frac{\partial\Phi}{\partial t} + f(t), \tag{3.96}$$

where $f(t)$ is an arbitrary function of time with the dimension of length.

3.7 Irrotational and Divergence-Free Flow

If the flow is both irrotational and divergence-free, we can represent the velocity vector both as the gradient of a scalar potential and as the curl of a vector potential. A scalar potential is a simpler quantity to deal with than a vector potential, which has three components. For this reason, we normally represent an irrotational and divergence-free vector field in terms of the gradient of a scalar potential, i.e.,

$$v_i = -\partial_i\Phi. \tag{3.97}$$

Since the divergence is zero, the potential satisfies Laplace's equation,

$$\partial_i\partial_i\Phi = \frac{\partial^2\Phi}{\partial x_i \partial x_i} = \nabla^2\Phi = 0. \tag{3.98}$$

3.8 Three-Dimensional Flow

Laplace's equation, which governs irrotational divergence-free flow is linear; we can construct new solutions by superimposing existing ones. We present in this section a number of elementary solutions to Laplace's equation, and combine them to construct new ones.

3.8.1 The Point Sink

A point sink is a point in three-dimensional space where a fixed amount of fluid is withdrawn from the system. We represent the velocity vector as the gradient of a velocity potential, Φ,

$$v_i = -\partial_i\Phi. \tag{3.99}$$

We use the minus sign so that the flow is in the direction of decreasing potential. We represent the discharge of the point sink as Q and apply the mass balance equation, i.e., the total flow through any sphere with the point sink at its center must be Q. If the radial component of flow is v_r then

$$Q = 4\pi r^2(-v_r), \tag{3.100}$$

where v_r is positive for flow away from the point sink. The velocity is equal to minus the gradient of the potential

$$v_r = -\frac{d\Phi}{dr} = -\frac{Q}{4\pi r^2}, \tag{3.101}$$

so that

$$\Phi = -\frac{Q}{4\pi}\frac{1}{r}. \tag{3.102}$$

If the point sink is at $\overset{0}{x}_i$, then

$$r = \sqrt{(x_i - \overset{0}{x}_i)(x_i - \overset{0}{x}_i)}. \tag{3.103}$$

For the sake of convenience, we introduce a function u,[3]

$$u = (x_i - \overset{0}{x}_i)(x_i - \overset{0}{x}_i) = r^2, \tag{3.104}$$

and obtain the following expression for v_i

$$v_i = -\partial_i\Phi = \frac{Q}{4\pi}\partial_i\frac{1}{u^{1/2}} = -\frac{Q}{8\pi}\frac{\partial_i u}{u^{3/2}}. \tag{3.105}$$

We compute the gradient of u as follows:

$$\partial_i u = (x_j - \overset{0}{x}_j)\partial_i(x_j - \overset{0}{x}_j) + \partial_i(x_j - \overset{0}{x}_j)(x_j - \overset{0}{x}_j) = 2(x_j - \overset{0}{x}_j)\delta_{ij} = 2(x_i - \overset{0}{x}_i), \tag{3.106}$$

so that

$$v_i = -\partial_i\Phi = -\frac{Q}{8\pi}\frac{1}{u^{3/2}}2(x_i - \overset{0}{x}_i) = -\frac{Q}{4\pi}\frac{x_i - \overset{0}{x}_i}{u^{3/2}}. \tag{3.107}$$

We verify that this velocity field is indeed divergence-free by contracting (3.107) with ∂_i which gives

$$\partial_i v_i = -\partial_i\partial_i\Phi = -\nabla^2\Phi = \frac{3Q}{8\pi}\frac{x_i - \overset{0}{x}_i}{u^{5/2}}\partial_i u - \frac{Q}{4\pi}\frac{\partial_i(x_i - \overset{0}{x}_i)}{u^{3/2}}, \tag{3.108}$$

or

$$\partial_i v_i = -\nabla^2\Phi = \frac{3Q}{8\pi}\frac{2(x_i - \overset{0}{x}_i)(x_i - \overset{0}{x}_i)}{u^{5/2}} - \frac{Q}{4\pi}\frac{\delta_{ii}}{u^{3/2}}. \tag{3.109}$$

[3]We obtain a useful formula for $\partial_i r$ from this. We have $\partial_i u = \partial_i r^2 = 2r\partial_i r = 2(x_i - \overset{0}{x}_i)$. If we set $r_i = x_i - \overset{0}{x}_i$, then $\partial_i r = r_i/r$.

We note that

$$\delta_{ii} = \delta_{11} + \delta_{22} + \delta_{33} = 3, \tag{3.110}$$

so that indeed:

$$\partial_i v_i = \frac{3Q}{4\pi}\frac{u}{u^{5/2}} - 3\frac{Q}{4\pi}\frac{1}{u^{3/2}} = 0. \tag{3.111}$$

We obtain the flow into the point-sink (the discharge per unit time) by multiplying the component of v_i normal to a sphere of radius r by the surface area of the sphere. The normal component of v_i is

$$v_i n_i = v_i \frac{x_i - \overset{0}{x}_i}{r} = -\frac{Q}{4\pi}\frac{x_i - \overset{0}{x}_i}{r^3}\frac{x_i - \overset{0}{x}_i}{r} = -\frac{Q}{4\pi}\frac{1}{r^2}, \tag{3.112}$$

which corresponds to (3.101).

3.8.2 The Dipole

A dipole is the limiting case of a source and a sink that are a distance $2d$ apart, where the distance d decreases as Q increases such that

$$\lim_{\substack{d \to 0 \\ Q \to \infty}} Qd = A, \tag{3.113}$$

where A is a constant, see Figure 3.7. The point source is at $\overset{0}{x}_i + de_i$ and the point sink at $\overset{0}{x}_i - de_i$ where e_i is a unit vector that points from the point sink to the point source.

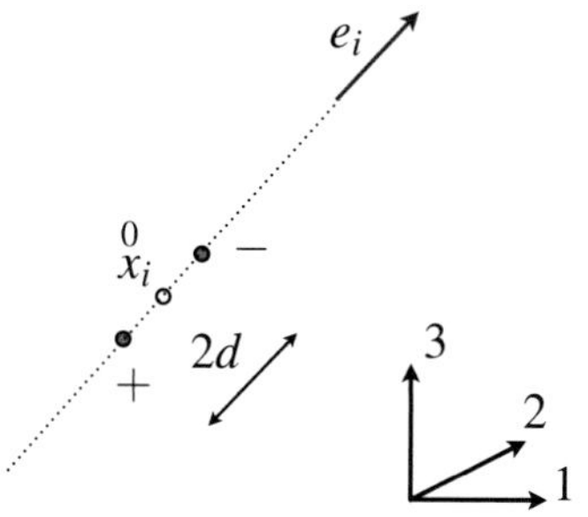

Figure 3.7: A dipole

The expression for the dipole is

$$\begin{aligned}\Phi &= \lim_{\substack{d \to 0 \\ Q \to \infty}} \frac{Q}{4\pi}[F(x_m, \overset{0}{x}_m + de_m) - F(x_i, \overset{0}{x}_m - de_m] \\ &= \lim_{\substack{d \to 0 \\ Q \to \infty}} \frac{2A}{4\pi}\frac{F(x_m, \overset{0}{x}_m + de_m) - F(x_i, \overset{0}{x}_m - de_m)}{2d},\end{aligned} \tag{3.114}$$

where

$$F(x_m, y_m) = \frac{1}{\sqrt{(x_m - y_m)(x_m - y_m)}}. \tag{3.115}$$

The function $F(x_m, y_m)$ is symmetrical in terms of x_m and y_m. The expression (3.114) represents the derivative of F in the direction e_i with respect to $\overset{0}{x}_i$. Although F is symmetrical with respect to x_i and y_i, the sign of the derivative with respect to y_i is opposite to that with respect to x_i.[4] Thus, the potential for the dipole is equal to minus the projection of the gradient of F onto the direction of e_i and we obtain

$$\Phi = -\frac{A}{2\pi} e_i \partial_i F. \tag{3.116}$$

The gradient of F is

$$\partial_i F = -\frac{r_i}{r^3}. \tag{3.117}$$

The expression for a dipole at $\overset{0}{x}_i$ with strength A is thus

$$\Phi = \frac{A}{2\pi} \frac{e_i r_i}{r^3} = \frac{A}{2\pi} \frac{e_i (x_i - \overset{0}{x}_i)}{[(x_j - \overset{0}{x}_j)(x_j - \overset{0}{x}_j)]^{3/2}}. \tag{3.118}$$

We present in small print an alternative derivation of the potential for a dipole, which consists of the limit of a source at $\overset{0}{x}_i + de_i$ and a sink at $\overset{0}{x}_i - de_i$ for the discharge approaching infinity as the source and the sink approach each other indefinitely. The expression for the velocity potential for a dipole is then, if the source is at $\overset{0}{x}_i + d_i$ and the sink at $\overset{0}{x}_i - d_i$

$$\Phi = -\lim_{\substack{d \to 0 \\ Q \to \infty}} \frac{Q}{4\pi} \left[\frac{1}{\sqrt{(x_i - \overset{0}{x}_i + de_i)(x_i - \overset{0}{x}_i + de_i)}} - \frac{1}{\sqrt{(x_i - \overset{0}{x}_i - de_i)(x_i - \overset{0}{x}_i - de_i)}} \right], \tag{3.119}$$

or

$$\Phi = -\lim_{\substack{d \to 0 \\ Q \to \infty}} \frac{Q}{4\pi} \left[\frac{1}{\sqrt{r^2 + 2de_i r_i + d^2}} - \frac{1}{\sqrt{(r^2 - 2de_i r_i + d^2)}} \right], \tag{3.120}$$

where

$$r_i = x_i - \overset{0}{x}_i \qquad r^2 = r_i r_i = (x_i - \overset{0}{x}_i)(x_i - \overset{0}{x}_i). \tag{3.121}$$

We rewrite (3.120):

$$\Phi = -\lim_{\substack{d \to 0 \\ Q \to \infty}} \frac{Q}{4\pi r} \left[\frac{1}{\sqrt{1 + (2de_i r_i + d^2)/r^2}} - \frac{1}{\sqrt{1 - (2de_i r_i - d^2)/r^2}} \right]. \tag{3.122}$$

The term $(1+x)^{-1/2}$, $x \to 0$, approaches $1 - \frac{1}{2}x$, as shown in the footnote[5]; we write (3.122) as

$$\Phi = -\lim_{\substack{d \to 0 \\ Q \to \infty}} \frac{Q}{4\pi r} \left[1 - \frac{2d_i r_i + d^2}{2r^2} - 1 - \frac{2d_i r_i - d^2}{2r^2} \right] = \lim_{\substack{d \to 0 \\ Q \to \infty}} \frac{Q}{4\pi r} \frac{2d_i r_i}{r^2}. \tag{3.123}$$

---

[4] The derivative $\partial/\partial y_i$ of $(x_m - y_m)(x_m - y_m)$ is $2(x_m - y_m)(-\partial y_i/\partial y_m)$ or $-2(x_i - y_i)$ whereas the derivative with respect to x_i is $2(x_i - y_i)$.

[5] To find the largest term, the one of order x, of $f = (1+x)^{1/2}$, we write $f = [(1 + \frac{1}{2}x)^2 - (\frac{1}{2}x)^2]^{1/2}$ and let $x \to 0$, so that $(x/2)^2$ can be neglected relative to $(1 + \frac{1}{2}x)^2$. Thus, $f \to 1 + \frac{1}{2}x$ for $x \to 0$, and $1/f \to 1/(1 + \frac{1}{2}x) = (1 - \frac{1}{2}x)/(1 - (\frac{1}{2}x)^2)$ which indeed approaches $1 - \frac{1}{2}x$ for $x \to 0$.

We write d_i as de_i, where e_i is a unit vector pointing in the direction of d_i,

$$\Phi = \lim_{\substack{d \to 0 \\ Q \to \infty}} \frac{Qd}{2\pi} \frac{r_i e_i}{r^3} = \frac{A}{2\pi} \frac{r_i e_i}{r^3}. \tag{3.124}$$

This result is identical to that obtained by differentiation, (3.118).

Note that it is not necessary to verify that this potential is harmonic, i.e., that the flow is divergence-free; we added two solutions that are harmonic, and the result will be harmonic as well.

3.8.3 Higher-Order Singularities

We may extend the process of letting point sinks (often called poles) approach one another as their discharges grow, to obtain expressions for higher order poles. It can be shown, in view of the analogy of the limiting process used above, that this amounts to taking higher order derivatives. Thus, a three-dimensional pole of order 3 can be written by analogy with (3.116) as

$$\Phi = -e_j \partial_j \Phi = \frac{A}{2\pi} \left[3 \frac{r_i \overset{1}{e}_i r_j \overset{2}{e}_j}{r^5} - \frac{\delta_{ij} \overset{1}{e}_i \overset{2}{e}_j}{r^3} \right] = \frac{A}{4\pi} \frac{\overset{1}{e}_i \overset{2}{e}_j}{r^5} [3 r_i r_j - \delta_{ij}]. \tag{3.125}$$

Note that the differentiation is with respect to $\overset{0}{x}_i$, which affects the sign of the derivative of r. The third-order pole has five degrees of freedom: two degrees of freedom each for the two different unit vectors $\overset{k}{e}_i$, plus the constant factor A.

The process of constructing higher order poles may be continued to obtain potentials for a series of poles, which can be used to create mathematical expressions to solve certain boundary-value problems. The pole of one order higher than the previous one will have three degrees of freedom added; each time a new unit vector is introduced along with a new scalar factor.

3.8.4 Other Harmonic Functions

The function $\Phi = e_j \partial_j \phi$ is harmonic if ϕ is harmonic and e_i is a constant vector:

$$\partial_i \Phi = \partial_i (e_j \partial_j \phi) = e_j \partial_i \partial_j \phi, \tag{3.126}$$

so that

$$\partial_i \partial_i \Phi = e_j \partial_i \partial_i \partial_j \phi = 0. \tag{3.127}$$

3.8.5 An Impermeable Sphere in a Field of Uniform Flow

The potential for a dipole by itself is not useful, but its combination with the potential for uniform flow results in the potential for an impermeable sphere inserted in a uniform flow field. The potential for uniform flow with velocity $\overset{0}{v}_i$ is

$$\Phi = -\overset{0}{v}_j x_j, \tag{3.128}$$

with the corresponding velocity field

$$v_i = -\partial_i \Phi = \overset{0}{v}_j \delta_{ij} = \overset{0}{v}_i. \tag{3.129}$$

We add the potential (3.124) for a dipole to this and choose the constant A so that the resulting potential yields a velocity field that meets the condition that no flow crosses the boundary of the sphere, which is centered at $\overset{0}{x}_i$ and has radius R. The potential is:

$$\Phi = -\overset{0}{v}_j x_j + \frac{A}{2\pi} \frac{r_j e_j}{r^3}. \tag{3.130}$$

The expression for the velocity v_i is:

$$v_i = -\partial_i \Phi = \overset{0}{v}_i - \frac{A}{2\pi} \left[\frac{\partial_i r_j e_j}{r^3} - 3 \frac{r_j e_j}{r^4} \partial_i r \right], \tag{3.131}$$

or, with

$$\partial_i r^2 = 2r \partial_i r = \partial_i (r_j r_j) = 2r_j \partial_i r_j = 2r_j \partial_{ij} = 2r_i \to \partial_i r = \frac{r_i}{r}, \tag{3.132}$$

(3.131) becomes

$$v_i = \overset{0}{v}_i - \frac{A}{2\pi} \left[\frac{\delta_{ij} e_j}{r^3} - 3 \frac{r_j e_j}{r^5} r_i \right], \tag{3.133}$$

or

$$v_i = \overset{0}{v}_i - \frac{A}{2\pi} \left[\frac{e_i}{r^3} - 3 \frac{r_j e_j r_i}{r^5} \right]. \tag{3.134}$$

The Boundary Condition

The condition at the boundary of the sphere is that the component of v_i normal to the sphere is zero. Since r_i is normal to the sphere, the dot product of v_i and r_i must be zero:

$$v_i r_i = 0 \qquad r = R. \tag{3.135}$$

We contract (3.134) with r_i:

$$v_i r_i = \overset{0}{v}_i r_i - \frac{A}{2\pi} \left[\frac{e_i r_i}{r^3} - 3 \frac{r_j e_j r_i r_i}{r^5} \right] \qquad r = R, \tag{3.136}$$

or, with $r_i r_i = r^2$,

$$v_i r_i = \overset{0}{v}_i r_i - \frac{A}{2\pi} \frac{-2e_i r_i}{R^3} = \left[\overset{0}{v}_i + \frac{A}{\pi R^3} e_i \right] r_i \qquad r = R. \tag{3.137}$$

This expression is zero only if the unit vector e_i points in the direction of v_i, so that

$$e_i = \frac{\overset{0}{v}_i}{\overset{0}{v}}. \tag{3.138}$$

and if

$$A = -\pi R^3 \overset{0}{v}, \tag{3.139}$$

We obtain the expression for the potential for an impermeable sphere in a field of uniform flow by substituting (3.139) for A into (3.130):

$$\Phi = -\overset{0}{v}\left[e_j x_j + \tfrac{1}{2} R^3 \frac{r_j e_j}{r^3}\right]. \tag{3.140}$$

We set $\overset{0}{v}_i = \overset{0}{v} e_i$ and $A = -\pi R^3 \overset{0}{v}$ in the expression for the velocity vector, (3.134),

$$v_i = \overset{0}{v}_i \left[1 + \tfrac{1}{2}\left(\frac{R}{r}\right)^3\right] - \tfrac{3}{2}\overset{0}{v}_j \frac{r_i r_j}{r^2}\left(\frac{R}{r}\right)^3, \tag{3.141}$$

where

$$\overset{0}{v} e_j = \overset{0}{v}_j. \tag{3.142}$$

Hydraulic Head and Pressure Head

The energy head is a constant for irrotational and divergence-free flow, which allows us to obtain expressions for both the pressure head and the hydraulic head ϕ:

$$H = \frac{p}{\rho g} + x_3 + \frac{v^2}{2g} = \phi + \frac{v^2}{2g} = H_0, \tag{3.143}$$

where:

$$\phi = \frac{p}{\rho g} + x_3. \tag{3.144}$$

The constant H_0 in (3.143) is some arbitrary reference value, obtained, for example, by setting the piezometric head to some fixed value at infinity. We use expression (3.141) for v_i to compute v^2,

$$v_i v_i = \overset{0}{v}{}^2 \left[1 + \tfrac{1}{2}\frac{R^3}{r^3}\right]^2 - 3\frac{(\overset{0}{v}_i r_i)^2 R^3}{r^5}\left[1 + \tfrac{1}{2}\frac{R^3}{r^3}\right] + \tfrac{9}{4}\left(\frac{r_j \overset{0}{v}_j R^3}{r^4}\right)^2, \tag{3.145}$$

or

$$v^2 = \overset{0}{v}{}^2 \left[1 + \tfrac{1}{2}\frac{R^3}{r^3}\right]^2 - 3\frac{(\overset{0}{v}_i r_i)^2 R^3}{r^5}\left[1 - \tfrac{1}{4}\frac{R^3}{r^3}\right]. \tag{3.146}$$

This equation is programmed in MATLAB to produce contours of constant pressure for the case of a velocity of 1 m/d and a sphere of radius $R = 100$ m, see Figure 3.8. The plot corresponds to the pressures on the sphere computed in the plane $z = 0$, with the sphere centered at the origin.

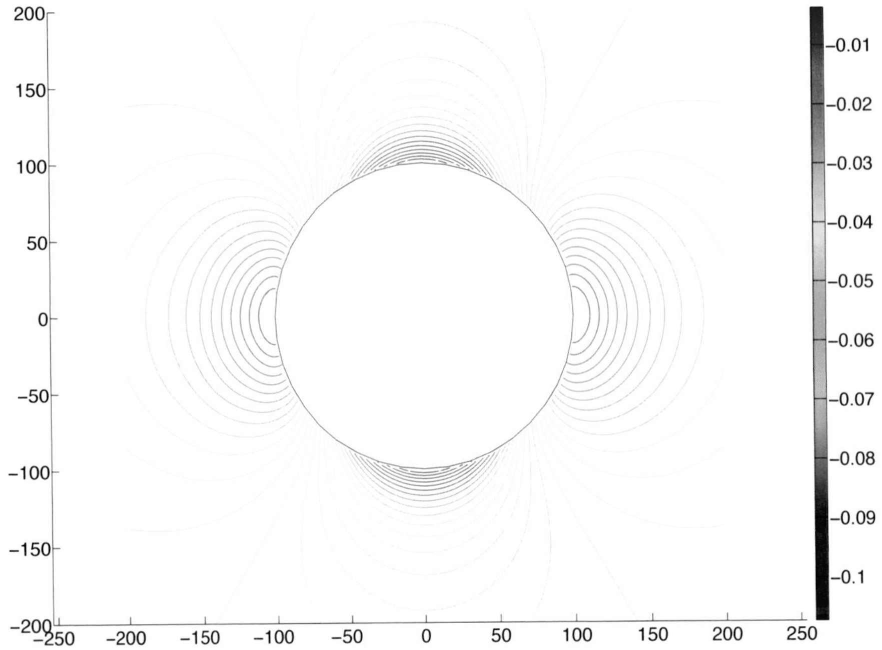

Figure 3.8: Pressures in the x_1,x_2-plane for a sphere of radius $R = 100$ m, in a velocity field of 1 m/d; the sphere is at the origin

3.8.6 The Line-Sink

A line-sink in three-dimensional space is a line of withdrawal at a constant rate in time, with the extraction rate taken as a constant, $\overset{0}{\sigma}$, along the line. We present a derivation that closely follows Duschek and Hochrainer [1970].

We obtain an expression for the potential for a line sink by integrating the potential for a point sink along a straight line from starting point $\overset{1}{x}_i$ to end point $\overset{2}{x}_i$, see figure 3.9. We use the local variables t_i, $\overset{0}{x}_i$, and h_i as:

$$t_i = x_i - \overset{0}{x}_i \tag{3.147}$$

$$\overset{0}{x}_i = \tfrac{1}{2}(\overset{1}{x}_i + \overset{2}{x}_i) \tag{3.148}$$

$$h_i = \tfrac{1}{2}\left(\overset{2}{x}_i - \overset{1}{x}_i\right), \tag{3.149}$$

and introduce the vectors u_i and v_i as

$$u_i = t_i - h_i, \tag{3.150}$$

$$v_i = t_i + h_i. \tag{3.151}$$

We use the potential for a point sink, (3.102), and write the integral of the point sink

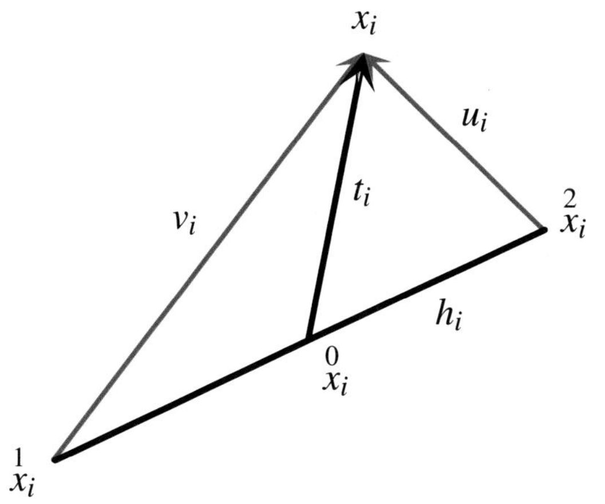

Figure 3.9: A line-sink in three-dimensional space

at p_i with discharge $Q = \sigma ds$ where σ is the extraction rate per unit length of line as follows,

$$\Phi = -\frac{1}{4\pi}\int_{\overset{1}{x}_i}^{\overset{2}{x}_i} \frac{\sigma}{\sqrt{(x_j - p_j)(x_j - p_j)}} ds, \tag{3.152}$$

where p_i represents an arbitrary point on the line,

$$p_i = \overset{0}{x}_i + \lambda h_i \qquad -1 \leq \lambda \leq 1. \tag{3.153}$$

We write the increment ds in terms of λ as $ds = hd\lambda$,

$$\Phi = -\frac{1}{4\pi}\int_{-1}^{1} \frac{h\sigma}{\sqrt{(x_j - \overset{0}{x}_j - \lambda h_j)(x_j - \overset{0}{x}_j - \lambda h_j)}} d\lambda. \tag{3.154}$$

We use the independent variable t_i defined in (3.147) and work out the term under the square root

$$\Phi = -\frac{1}{4\pi}\int_{-1}^{1} \frac{h\sigma}{\sqrt{t^2 - 2\lambda h_j t_j + \lambda^2 h^2}} d\lambda = -\frac{h\sigma}{4\pi}\int_{-1}^{1} \frac{1}{h\sqrt{\lambda^2 - 2\lambda \tau_j e_j + \tau^2}} d\lambda \tag{3.155}$$

where

$$\tau_i = \frac{t_i}{h}, \tag{3.156}$$

and

$$e_i = \frac{h_i}{h}. \tag{3.157}$$

The integral is in most integral tables:

$$\Phi = -\frac{\sigma}{4\pi}\int_{-1}^{1}\frac{1}{\sqrt{\lambda^2 - 2\lambda\tau_j e_j + \tau^2}}d\lambda$$
$$= -\frac{\sigma}{4\pi}\ln\left[\sqrt{\lambda^2 - 2\tau_j e_j\lambda + \tau^2} + \lambda - \tau_j e_j\right]\Bigg|_{-1}^{1}, \tag{3.158}$$

or

$$\Phi = -\frac{\sigma}{4\pi}\ln\frac{\sqrt{\tau^2 - 2\tau_j e_j + 1} + 1 - \tau_j e_j}{\sqrt{\tau^2 + 2\tau_j e_j + 1} - 1 - \tau_j e_j}. \tag{3.159}$$

We introduce dimensionless variables U_i and V_i as

$$\begin{aligned} U_i &= \tau_i - e_i = \frac{t_i - h_i}{h} \\ V_i &= \tau_i + e_i = \frac{t_i + h_i}{h}. \end{aligned} \tag{3.160}$$

The following identities turn out to be useful:

$$\begin{aligned} U_i e_i &= e_i\tau_i - 1 \qquad & U^2 &= (\tau_i - e_i)(\tau_i - e_i) = \tau^2 - 2\tau_i e_i + 1 \\ V_i e_i &= e_i\tau_i + 1 \qquad & V^2 &= (\tau_i + e_i)(\tau_i + e_i) = \tau^2 + 2\tau_i e_i + 1. \end{aligned} \tag{3.161}$$

We use these new variables to rewrite (3.159) as

$$\Phi = -\frac{\sigma}{4\pi}\ln\frac{U - U_i e_i}{V - V_i e_i}. \tag{3.162}$$

Equipotentials

We obtain equations for the equipotentials, i.e., the surfaces defined by constant values of Φ by setting Φ equal to a constant in (3.162),

$$\frac{U - U_i e_i}{V - V_i e_i} = \mathrm{e}^{-4\pi\Phi/\sigma}. \tag{3.163}$$

We write e_i as

$$e_i = -\tfrac{1}{2}(U_i - V_i), \tag{3.164}$$

so that

$$\begin{aligned} U_i e_i &= -\tfrac{1}{2}U_i(U_i - V_i) = -\tfrac{1}{2}(U_i - V_i)\left[\tfrac{1}{2}(U_i + V_i) + \tfrac{1}{2}(U_i - V_i)\right] \\ &= -\tfrac{1}{4}(U^2 - V^2) - 1, \end{aligned} \tag{3.165}$$

and

$$\begin{aligned} V_i e_i &= -\tfrac{1}{2}V_i(U_i - V_i) = -\tfrac{1}{2}(U_i - V_i)\left[\tfrac{1}{2}(U_i + V_i) - \tfrac{1}{2}(U_i - V_i)\right] \\ &= -\tfrac{1}{4}(U^2 - V^2) + 1. \end{aligned} \tag{3.166}$$

We use the latter two expressions to rewrite (3.163):

$$\frac{U - U_i e_i}{V - V_i e_i} = \frac{\frac{1}{2}(U+V) + \frac{1}{2}(U-V) + \frac{1}{4}(U-V)(U+V) + 1}{\frac{1}{2}(U+V) - \frac{1}{2}(U-V) + \frac{1}{4}(U-V)(U+V) - 1}, \tag{3.167}$$

or

$$\frac{U - U_i e_i}{V - V_i e_i} = \frac{\frac{1}{2}(U+V)[\frac{1}{2}(U-V)+1] + [\frac{1}{2}(U-V)+1]}{\frac{1}{2}(U+V)[\frac{1}{2}(U-V)+1] - [\frac{1}{2}(U-V)+1]}. \tag{3.168}$$

We exclude the special case that $U - V = -2$ so that we can divide by the common factor:

$$\frac{U - U_i e_i}{V - V_i e_i} = \frac{\frac{1}{2}(U+V)+1}{\frac{1}{2}(U+V)-1}, \tag{3.169}$$

and set Φ equal to a constant in (3.163) to obtain the equation for the equipotentials:

$$\frac{\frac{1}{2}(U+V)+1}{\frac{1}{2}(U+V)-1} = \mathrm{e}^{-4\pi\Phi_0/\sigma}. \tag{3.170}$$

This equation represents a family of rotational ellipsoids with their common axes coinciding with the line element, and with the foci of the ellipsoids at the end points of the line element. We use (3.169) to write an alternative expression for the potential (3.162),

$$\Phi = -\frac{\sigma}{4\pi} \ln \frac{\frac{1}{2}(U+V)+1}{\frac{1}{2}(U+V)-1}. \tag{3.171}$$

Singular Behavior

We observe from (3.163) that the potential is infinite when either $U_i e_i = U$ or $V_i e_i = V$, i.e., when the length of the vectors from either end point to point x_i equals its projection on the line sink, which happens only if x_i is a point on the line sink; the potential approaches negative infinity as x_i approaches the line. This is to be expected; a finite amount of fluid (or heat) can be extracted only if the line sink has a finite size.

Contours of constant potential are shown in Figure 3.10 for a vertical section through the line-sink.

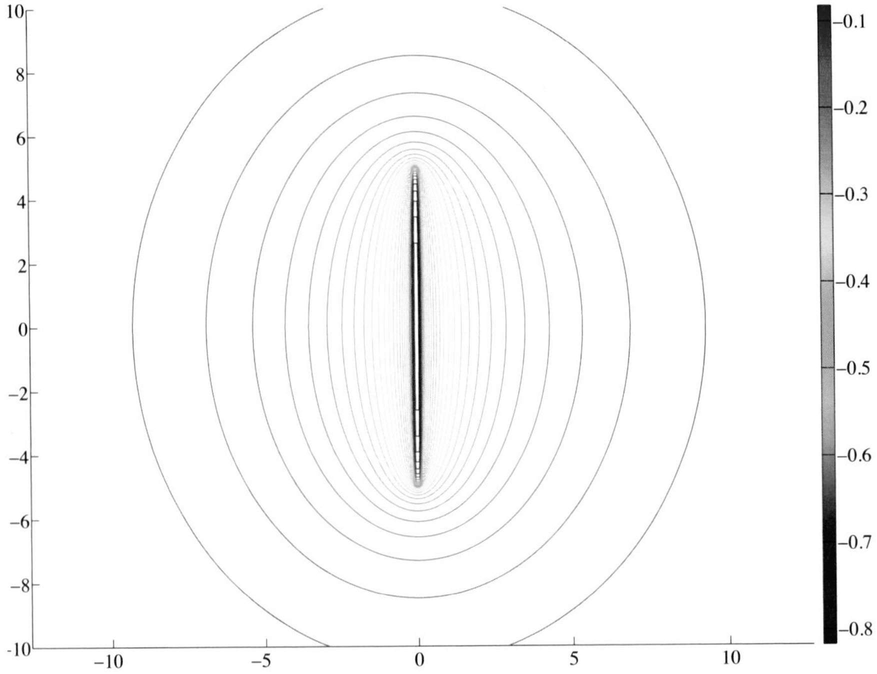

Figure 3.10: Linesink

Chapter 4

Integral Theorems

4.1 Introduction

Integral theorems relate integrals over surfaces to integrals over their boundaries, and integrals over volumes to integrals over their bounding surfaces. Stokes's integral theorem applies to a surface, the divergence theorem, which is a special case of Gauss's integral theorem, concerns a volume, and the integral theorems of Green also concern a volume.

The integral theorem of Stokes relates the integral of the curl of a vector over a surface to the integral of the tangential component of that vector over the boundary of the surface. The divergence theorem relates the integral of the divergence of a vector over a volume to the integral of the normal component of that vector over the bounding surface of the volume. The integral theorem of Green provides a relationship between the Laplacians of two unrelated functions, and is the basis for boundary integral methods.

We follow the analysis presented by Duschek and Hochrainer [1970] to a large extent, but replace mathematical derivations by physical ones where possible, to make the explanations more easily understood for engineering students and professionals. The reader is referred to Duschek and Hochrainer [1970] for rigorous derivations.

4.2 The Integral Theorem of Stokes

Following Duschek and Hochrainer [1970], we define an incremental surface element df_i, as a vector that points normal to the surface and of a magnitude equal to the incremental surface element $d\mathscr{S}$. We create this vector by considering a parametric representation of the surface, $x_i = x_i(u,v)$, where the parameters u and v are chosen such that constant values of u and v each define curves that lie on the surface. The vectors $\partial x_i/\partial u$ and $\partial x_i/\partial v$ represent tangent vectors to the curves, in much the same way as the tangent vector $T_i = dx_i/ds$ to a curve is defined by $x_i = x_i(s)$. We choose the parameters u and v to be Cartesian coordinates. This definition of u and v implies restrictions, for example that the correspondence between u and v and the curves on $\mathscr{S}$ is unique and reversible, and may make it necessary to divide the surface into parts.

O. D. L. Strack, *Applications of Vector Analysis and Complex Variables in Engineering*, https://doi.org/10.1007/978-3-030-41168-8_4

This is the case, for example if we consider the surface of a sphere, which must be split into two semi-spheres, each with its own set of u, v parameters.

We use this parametric representation of the surface to define the elementary surface element df_i, which is a vector normal to the surface, as:

$$df_i = \varepsilon_{ijk} \frac{\partial x_i}{\partial u} \frac{\partial x_j}{\partial v} du dv. \tag{4.1}$$

We consider the following vector integral over the surface $\mathscr{S}$:

$$I_k = \iint\limits_{\mathscr{S}} \varepsilon_{ijk} \frac{\partial A}{\partial x_j} df_i, \tag{4.2}$$

express df_i according to (4.1):

$$I_k = \iint\limits_{\mathscr{S}} \varepsilon_{ijk} \varepsilon_{imn} \frac{\partial A}{\partial x_j} \frac{\partial x_m}{\partial u} \frac{\partial x_n}{\partial v} du dv, \tag{4.3}$$

and apply the relationship between the epsilon and delta tensors:

$$\begin{aligned} I_k &= \iint\limits_{\mathscr{S}} (\delta_{jm}\delta_{kn} - \delta_{jn}\delta_{km}) \frac{\partial A}{\partial x_j} \frac{\partial x_m}{\partial u} \frac{\partial x_n}{\partial v} du dv \\ &= \iint\limits_{\mathscr{S}} \left(\frac{\partial A}{\partial x_j} \frac{\partial x_j}{\partial u} \frac{\partial x_k}{\partial v} - \frac{\partial A}{\partial x_j} \frac{\partial x_j}{\partial v} \frac{\partial x_k}{\partial u} \right) du dv, \end{aligned} \tag{4.4}$$

or

$$I_k = \iint\limits_{\mathscr{S}} \left(\frac{\partial A}{\partial u} \frac{\partial x_k}{\partial v} - \frac{\partial A}{\partial v} \frac{\partial x_k}{\partial u} \right) du dv. \tag{4.5}$$

We choose the parameters u and v, following Duschek and Hochrainer [1970], as special Cartesian coordinates for each of the three components of the vector integral. For $k = 1$, we choose $u = x_1$ and $v = x_2$, so that $\mathscr{S}$ is represented by $x_3 = x_3(x_1, x_2)$. This gives, with $\partial x_1 / \partial x_2 = 0$,

$$I_1 = - \iint\limits_{\mathscr{S}} \frac{\partial A}{\partial x_2} dx_1 dx_2. \tag{4.6}$$

We integrate with respect to x_2 along the curves $x_1 =$ constant, from the one side, $\underset{s}{\mathscr{B}}$, of the boundary to the other, $\underset{e}{\mathscr{B}}$, as illustrated in Figure 4.1,

$$I_1 = - \int\limits_{\underset{s}{\mathscr{B}}}^{\underset{e}{\mathscr{B}}} A dx_1 = - \int \left(\underset{e}{A} - \underset{s}{A} \right) dx_1. \tag{4.7}$$

The sign of the integral along $\underset{e}{\mathscr{B}}$ is opposite to that along $\underset{s}{\mathscr{B}}$ as the directions of

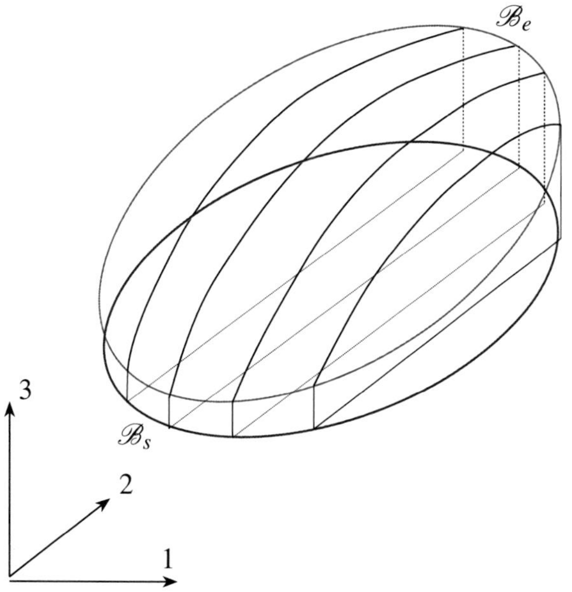

Figure 4.1: Integration over the surface along curves x_1 =constant; adapted from Duschek and Hochrainer [1970]

integration are opposite. We combine the two integrals into a single one:

$$I_1 = \int_{\mathscr{B}} A dx_1. \tag{4.8}$$

We carry out the integrations for the remaining two terms of I_k in a similar manner, choosing $u = x_2$ and $v = x_3$ for I_2 and $u = x_3$ and $v = x_1$ for I_3, so that

$$\begin{aligned} I_2 &= \int_{\mathscr{B}} A dx_2 \\ I_3 &= \int_{\mathscr{B}} A dx_3. \end{aligned} \tag{4.9}$$

We combine the three integrals into a single vector:

$$I_k = \int_{\mathscr{B}} A dx_k, \tag{4.10}$$

and obtain Stokes's integral theorem by replacing I_k by (4.2):

$$\iint_{\mathscr{S}} \varepsilon_{ijk} \frac{\partial A}{\partial x_j} df_i = \int_{\mathscr{B}} A dx_k. \tag{4.11}$$

If the conditions we imposed on the parametric representation of the surface, namely that there is a one-to-one correspondence between the Cartesian coordinates u, v, and points on the surface are not met, we divide the surface into sub-surfaces such that each satisfies the conditions. The resulting boundaries of the sub domain have sections in common; the individual integrals along common sections are in opposite directions for each sub-area, and cancel. The remaining integral is along the outer boundary of the surface.

4.2.1 Stokes's Integral Theorem Applied to a Vector Potential

The practical application of Stokes's integral theorem concerns the case that a vector field v_i is rotational, but divergence-free, and is represented as the curl of a vector potential W_k:

$$v_k = -\varepsilon_{ijk}\partial_i W_j \tag{4.12}$$

We replace A in (4.11) by the vector W_k:

$$\iint\limits_{\mathscr{S}} v_i df_i = -\iint\limits_{\mathscr{S}} \varepsilon_{ijk}\partial_j W_k df_i = -\int\limits_{\mathscr{B}} W_k dx_k \tag{4.13}$$

This theorem states that the integral of the normal projection of a vector v_i over a surface $\mathscr{S}$ is equal to the integral along the boundary $\mathscr{B}$ of $\mathscr{S}$ of the tangential projection of the vector potential of v_i. The integral of the normal component of v_i over $\mathscr{S}$ represents the flow through the surface. Evidently, this flow is not affected by the particular shape of the surface, but only by its boundary, provided that the vector field is divergence-free. The vector potential is useful for computing flow rates through surfaces of arbitrary shape.

4.3 The Divergence Theorem

The divergence theorem is a special case of a more general theorem, known as Gauss' integral theorem, and states that the integral of the divergence, $\partial_i v_i$, of a vector v_i over a volume $\mathscr{V}$ is equal to the integral of the normal component of v_i over the surface $\mathscr{S}$ that bounds $\mathscr{V}$. This theorem is merely a statement of the principle of continuity of flow and takes the form:

$$\iiint\limits_{\mathscr{V}} \partial_i v_i d\mathscr{V} = \iint\limits_{\mathscr{S}} v_i df_i. \tag{4.14}$$

4.4 Green's Integral Theorem

The divergence theorem can be used to obtain Green's integral theorems. We apply the divergence theorem to a function FG_i,

$$\iiint\limits_{\mathscr{V}} \partial_i(FG_i)d\mathscr{V} = \int\limits_{\mathscr{S}} FG_i df_i. \tag{4.15}$$

We replace the vector function G_i by $\partial_i G$ and obtain Green's first identity:

$$\iiint_{\mathscr{V}} \partial_i (F \partial_i G) d\mathscr{V} = \iiint_{\mathscr{V}} (\partial_i G \partial_i F + F \partial_i \partial_i G) \, d\mathscr{V} = \iint_{\mathscr{S}} F \partial_i G df_i, \tag{4.16}$$

and apply this to $G\partial_i F$ instead of $F\partial_i G$:

$$\iiint_{\mathscr{V}} \partial_i (G \partial_i F) d\mathscr{V} = \iiint_{\mathscr{V}} (\partial_i G \partial_i F + G \partial_i \partial_i F) \, d\mathscr{V} = \iint_{\mathscr{S}} G \partial_i F df_i. \tag{4.17}$$

We subtract the latter two equations to obtain Green's second identity, replacing $\partial_i \partial_i$ by ∇^2:

$$\iiint_{\mathscr{V}} \left(F \nabla^2 G - G \nabla^2 F\right) d\mathscr{V} = \iint_{\mathscr{S}} (F \partial_i G - G \partial_i F) \, df_i. \tag{4.18}$$

4.5 Boundary Integral Equations

Green's second identity is the basis for a method for solving boundary-value problems, known as the boundary integral equation method, BIEM. We make a special choice for the function G:

$$G = \frac{1}{4\pi} \frac{1}{\sqrt{(x_i - p_i)(x_i - p_i)}} \tag{4.19}$$

This function is the velocity potential for a point sink of unit discharge at p_i. This function is singular at p_i; we exclude this singularity from the surface $\mathscr{S}$ by connecting it via a cylinder , $\mathscr{C}$, of infinitesimal radius with a sphere, $\mathscr{P}$, also of infinitesimal radius, around point p_i. We apply Green's second identity to this surface, and restrict the analysis to functions G that satisfy Laplace's equation, i.e., $\nabla^2 G = 0$, so that the volume integral vanishes:

$$\begin{aligned}
0 = & \frac{1}{4\pi} \iint_{\mathscr{S}} \left\{ F \partial_i \left(\frac{1}{\sqrt{(x_i - p_i)(x_i - p_i)}} \right) - \frac{1}{\sqrt{(x_i - p_i)(x_i - p_i)}} \partial_i F \right\} df_i \\
& + \frac{1}{4\pi} \iint_{\mathscr{C}} \left\{ F \partial_i \left(\frac{1}{\sqrt{(x_i - p_i)(x_i - p_i)}} \right) - \frac{1}{\sqrt{(x_i - p_i)(x_i - p_i)}} \partial_i F \right\} df_i \\
& + \frac{1}{4\pi} \iint_{\mathscr{P}} \left\{ F \partial_i \left(\frac{1}{\sqrt{(x_i - p_i)(x_i - p_i)}} \right) - \frac{1}{\sqrt{(x_i - p_i)(x_i - p_i)}} \partial_i F \right\} df_i.
\end{aligned} \tag{4.20}$$

If we let the radius of the cylinder reduce to zero, the integral along $\mathscr{C}$ vanishes, because the functions are continuous. We take the limit of the third integral as the radius of the sphere around p_i approaches zero. We define $r_i = x_i - p_i$ and carry out the differentiation with respect to x_i:

$$\partial_i \frac{1}{\sqrt{r_k r_k}} = -\frac{r_i}{r^3}. \tag{4.21}$$

We take the limit for $r \to 0$ of the integral over $\mathscr{P}$:

$$\frac{1}{4\pi} \lim_{x_i \to p_i} \iiint\limits_{\mathscr{P}} \left[\frac{F(x_j) r_i}{r^3} - \partial_i F \frac{1}{r} \right] d f_i. \tag{4.22}$$

The volume element $d f_I$ is a vector normal to the surface, i.e., in the direction of r_i, and r_i/r is a unit vector normal to the sphere. In the limit for $r \to 0$, $F(x_i)$ approaches $F(p_i)$, and the first integral reduces to the surface area of the sphere, $4\pi r^2$, multiplied by $-F(p_i)/r^2$, and thus approaches $-4\pi F(p_i)$ for $r \to 0$. The second integral vanishes, because $d f_i$ is of order r^2, so that the ratio $d f_i / r$ vanishes for $r \to 0$. The result is:

$$\frac{1}{4\pi} \iint\limits_{\mathscr{P}} \left\{ F \partial_i \left(\frac{1}{\sqrt{(x_j - p_j)(x_j - p_j)}} \right) - \frac{1}{\sqrt{(x_j - p_j)(x_j - p_j)}} \partial_i F \right\} d f_i = -F(p_i), \tag{4.23}$$

and we obtain with (4.20) and (4.21)

$$F(p_i) = -\frac{1}{4\pi} \iint\limits_{\mathscr{S}} \left\{ F \frac{(x_j - p_j)}{[(x_i - p_j)(x_i - p_j)]^{3/2}} + \frac{1}{\sqrt{(x_j - p_j)(x_i - p_j)}} \partial_i F \right\} d f_i. \tag{4.24}$$

The function G is often called the Green's function, but it is actually Green's function for the infinite domain. The Green's function in a strict sense of its definition is a function that behaves as a point sink of unit discharge at p_i, but is constructed in such a way that it satisfies the special boundary condition $G = 0$ along the surface $\mathscr{S}$. The second term inside the parentheses to the right of the equal sign in (4.18) vanishes, and, with the function F known, the integral can be evaluated. If the function G does not vanish along $\mathscr{S}$, both F and its gradient along the boundary are present in the integral. As explained in Chapter 8, only one boundary condition can be applied for Laplace's equation; the boundary integral thus contains one given function, and one unknown one.

The boundary integral equation method, BIEM, is based on approximating F, usually as a polynomial. The unknowns in this polynomial are solved for by applying (4.24) along the boundary.

Chapter 5

Coordinate Transformations: Definitions of Vectors and Tensors

We consider the formal definitions of vectors and tensors, and the equations for their components in a transformed coordinate system. We restrict the analysis to straight coordinates. We first consider Cartesian coordinates and afterward non-Cartesian coordinates.

5.1 Coordinate Transformation

We consider a Cartesian system $\mathscr{B}$ with coordinate axes $(\tilde{1},\tilde{2},\tilde{3})$, representing a point in this system as $\tilde{x}_i$. The origin of $\mathscr{B}$ in terms of the original coordinate system $\mathscr{A}$ (1,2,3) is given by the coordinates b_j, see Figure 5.1. Both coordinate systems are Cartesian, and the scales in both systems are the same. The vectors $\underset{m}{\tau}^i$, defined as

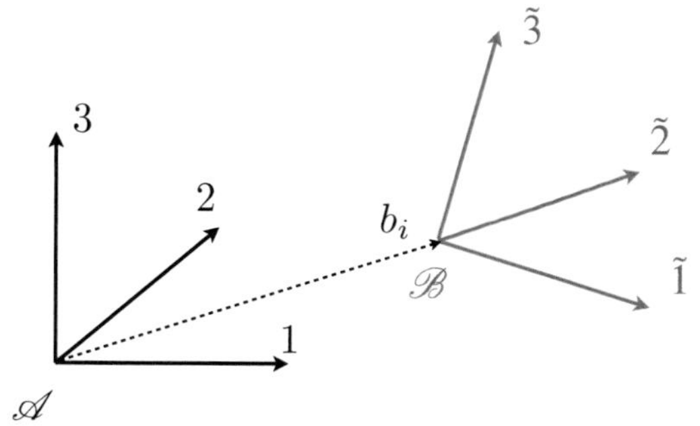

Figure 5.1: Transformation of Cartesian system $\mathscr{A}$ into Cartesian system $\mathscr{B}$

O. D. L. Strack, *Applications of Vector Analysis and Complex Variables in Engineering*, https://doi.org/10.1007/978-3-030-41168-8_5

$$\underset{m}{\tau}^{i} = \frac{\partial x_i}{\partial \tilde{x}_m} \qquad i = 1,2,3 \qquad m = 1,2,3, \tag{5.1}$$

are unit vectors tangent to the coordinate directions $\tilde{i}$ (compare (2.5)) with $\tilde{x}_m$ replacing the arc length along the curve, s. We define similar vectors that are tangent to the coordinates in system $\mathscr{A}$ in terms of the coordinates in system $\mathscr{B}$. We call these unit tangent vectors $\overset{m}{\gamma}_i$:

$$\overset{m}{\gamma}_i = \frac{\partial \tilde{x}_m}{\partial x_i}. \tag{5.2}$$

We express the original coordinates of a point x_i in terms of the new coordinates of that same point, and vice-versa, by integrating the latter two equations:

$$x_i = \underset{m}{\tau}^{i} \tilde{x}_m + b_i, \tag{5.3}$$

where b_i represents the origin of $\mathscr{B}$ in terms of x_i, and

$$\tilde{x}_m = \overset{m}{\gamma}_i x_i + \tilde{b}_m, \tag{5.4}$$

where $\tilde{b}_m$ represents the origin of the original system $\mathscr{A}$ in terms of the new coordinates.

The vectors $\underset{m}{\tau}^{i}$ are mutually orthogonal unit vectors,

$$\underset{m}{\tau}^{i} \underset{n}{\tau}^{i} = \delta_{mn}. \tag{5.5}$$

A similar equation holds for the unit vectors $\overset{m}{\gamma}_i$,

$$\overset{m}{\gamma}_i \overset{n}{\gamma}_i = \delta_{mn}. \tag{5.6}$$

We obtain a relationship between the vectors $\underset{m}{\tau}^{i}$ and $\overset{m}{\gamma}_i$ from (5.1) and (5.2)

$$\underset{m}{\tau}^{i} \overset{n}{\gamma}_i = \frac{\partial x_i}{\partial \tilde{x}_m} \frac{\partial \tilde{x}_n}{\partial x_i} = \frac{\partial \tilde{x}_n}{\partial \tilde{x}_m} = \delta_{mn}. \tag{5.7}$$

The moduli of $\underset{m}{\tau}^{i}$ and of $\overset{m}{\gamma}_i$ are 1; we combine this with the latter equation:

$$\underset{m}{\tau}^{i} = \overset{m}{\gamma}_i. \tag{5.8}$$

We introduce a matrix a_{ij},

$$a_{im} = \underset{m}{\tau}^{i} = \overset{m}{\gamma}_i = \frac{\partial x_i}{\partial \tilde{x}_m} = \frac{\partial \tilde{x}_m}{\partial x_i}, \tag{5.9}$$

and write (5.3) and (5.4) as

$$x_i = a_{ij} \tilde{x}_j + b_i, \tag{5.10}$$

and

$$\tilde{x}_i = a_{ji}x_j + \tilde{b}_i. \tag{5.11}$$

We observe from (5.9) that

$$a_{im}a_{in} = \frac{\partial x_i}{\partial \tilde{x}_m}\frac{\partial \tilde{x}_n}{\partial x_i} = \delta_{mn}, \tag{5.12}$$

and that

$$a_{im}a_{jm} = \frac{\partial \tilde{x}_m}{\partial x_i}\frac{\partial x_j}{\partial \tilde{x}_m} = \delta_{ij}. \tag{5.13}$$

5.2 Definition of Vectors and Tensors

A vector is a mathematical representation of some quantity that has both an orientation and a magnitude; it usually represents a physical quantity, such as the velocity of a water particle. A vector does not change its orientation and magnitude in space as a result of a coordinate transformation, which enables us to establish equations to determine if a set of three expressions represents a vector.

The differences between the coordinates of one point in space and another are the components of a vector; its length and position are invariant with respect to a coordinate transformation. Accordingly we introduce a vector A_i as

$$A_i = y_i - x_i \qquad i = 1,2,3, \tag{5.14}$$

where y_i and x_i are the coordinates of two arbitrary points in space. We use (5.11) to transform A_i into $\tilde{A}_i$:

$$\tilde{A}_i = \tilde{y}_i - \tilde{x}_i = a_{ji}(y_j - x_j) + \tilde{b}_i - \tilde{b}_i, \tag{5.15}$$

or

$$\tilde{A}_i = a_{ji}A_j. \tag{5.16}$$

We obtain the inverse transformation the same way from (5.10):

$$A_i = a_{ij}\tilde{A}_j. \tag{5.17}$$

The first index in a_{ij} corresponds to A_i in both equations. Equations (5.16) and (5.17) are the transformation formulas for a vector; any set of three components that transforms according to (5.16) and (5.17) is a vector in the sense defined above and represents a quantity that will not change due to a coordinate transformation.

5.2.1 Tensors of the Second Rank

A tensor of the second rank has nine components in three-dimensional space and has two indices, each ranging from 1 to 3. A tensor is an operator that produces a vector upon being applied to another vector. Thus, if the tensor A_{ij} operates on a vector X_j, the result is a new vector Y_i:

$$Y_i = A_{ij}X_j. \tag{5.18}$$

We obtain the transformation rules for a tensor of the second rank from those of a vector. We have, from (5.16)

$$\tilde{Y}_i = a_{ji} Y_j = a_{ji} A_{jm} X_m. \tag{5.19}$$

We use (5.17) to express X_m in terms of $\tilde{X}_m$;

$$\tilde{Y}_i = a_{ji} A_{jm} a_{mk} \tilde{X}_k = a_{ji} a_{mk} A_{jm} \tilde{X}_k. \tag{5.20}$$

We have, by definition

$$\tilde{Y}_i = \tilde{A}_{ij} \tilde{X}_j, \tag{5.21}$$

so that

$$\tilde{A}_{mn} = a_{im} a_{jn} A_{ij}. \tag{5.22}$$

The first index in the matrices a_{im} and a_{jn} refers to the original tensor components; the second index to those of the transformed one.

We obtain the inverse transformation in a similar way:

$$A_{mn} = a_{mi} a_{nj} \tilde{A}_{ij}. \tag{5.23}$$

Again, the first index in each transformation matrix refers to the original tensor.

A special tensor of the second rank is the Kronecker delta; its transformation is:

$$\tilde{\delta}_{mn} = a_{im} a_{jn} \delta_{ij} = a_{im} a_{in}, \tag{5.24}$$

or, with (5.13)

$$\tilde{\delta}_{mn} = \delta_{mn}. \tag{5.25}$$

The Kronecker delta is invariant with respect to a Cartesian coordinate transformation.

5.2.2 The Gradient Tensor

Consider a vector field A_i. We apply the operator ∂_i to this vector field to obtain a tensor, called the gradient tensor,

$$A_{ij} = \partial_i A_j \tag{5.26}$$

Problem 5.1

Demonstrate that $\partial_i A_j$ is a tensor of the second rank.

5.2.3 Principal Directions

A tensor operating on a vector X_i produces a new vector Y_i,

$$Y_i = A_{ij} X_j. \tag{5.27}$$

We choose the direction of the vector X_i such that the vector Y_i is collinear with X_i; the direction of this vector is the principal direction of the tensor. For this case, Y_i equals λX_i, where λ is some scalar quantity,

$$Y_i = \lambda X_i = A_{ij} X_j, \tag{5.28}$$

or

$$A_{ij}X_j - \lambda X_i = (A_{ij} - \lambda \delta_{ij})X_j = 0. \tag{5.29}$$

This represents a system of three homogeneous equations for the components of X_i, which has non-trivial solutions only if the determinant of the system is zero, i.e., if

$$\det(A_{ij} - \lambda \delta_{ij}) = \begin{vmatrix} A_{11} - \lambda & A_{12} & A_{13} \\ A_{21} & A_{22} - \lambda & A_{23} \\ A_{31} & A_{32} & A_{33} - \lambda \end{vmatrix} = 0. \tag{5.30}$$

This third-order equation is the characteristic equation of the tensor A_{ij}; its roots are the principal values of the tensor. The corresponding three directions of X_i are the characteristic directions, or principal directions, of the tensor. The homogenous equations can be divided by a fixed constant. If we choose for this constant the magnitude (modulus) of X_i, the result will be a unit vector, fully determined by two values. The principal values and directions are also known as the eigenvalues and eigenvectors of the tensor.

5.3 Non-Cartesian Transformations with Straight Axes

We consider straight transformed coordinates $\tilde{1},\tilde{2},\tilde{3}$ that are not orthogonal. That the coordinate axes are straight is an important restriction; differentiation along curves adds terms to derivatives, as we saw when establishing equations for the acceleration; the additional terms are the centrifugal acceleration, which is due entirely to the curvature of the path. Since we make restrictions to straight axes, such additional terms do not appear.

The definitions of the base vectors $\underset{m}{\tau}^i$ and $\overset{m}{\gamma}_i$ remain as before:

$$\underset{m}{\tau}^i = \frac{\partial x_i}{\partial \tilde{x}_m}, \tag{5.31}$$

and

$$\overset{m}{\gamma}_i = \frac{\partial \tilde{x}_m}{\partial x_i}. \tag{5.32}$$

The base vectors $\underset{m}{\tau}^i$ represent the tangent vectors to the coordinate axes $\tilde{1},\tilde{2},\tilde{3}$, but are not orthogonal. We interpret the base vectors $\overset{m}{\gamma}_i$ now in a different manner; we view them as the gradients of the functions $\tilde{x}_m(x_1,x_2,x_3)$, i.e., the base vector $\overset{1}{\gamma}_i$ points in the direction of the gradient of the function $\tilde{x}_1(x_1,x_2,x_3)$. The planes $\tilde{x}_1$=constant are, by definition, parallel to the $\tilde{2},\tilde{3}$-plane and the base vector $\overset{1}{\gamma}_i$ is normal to the $\tilde{2},\tilde{3}$-plane. The three base vectors $\overset{m}{\gamma}_i$, m=1,2,3 are thus orthogonal to the three planes $\tilde{2},\tilde{3}$, $\tilde{1},\tilde{3}$, and $\tilde{1},\tilde{2}$, respectively, see Figure 5.2.

An important difference between Cartesian and non-Cartesian coordinate systems is that the base vectors cannot all be unit vectors. We have the following relationships between the two sets of base vectors

$$\underset{m}{\tau}^i \overset{n}{\gamma}_i = \frac{\partial x_i}{\partial \tilde{x}_m} \frac{\partial \tilde{x}_n}{\partial x_i} = \delta_{mn}, \tag{5.33}$$

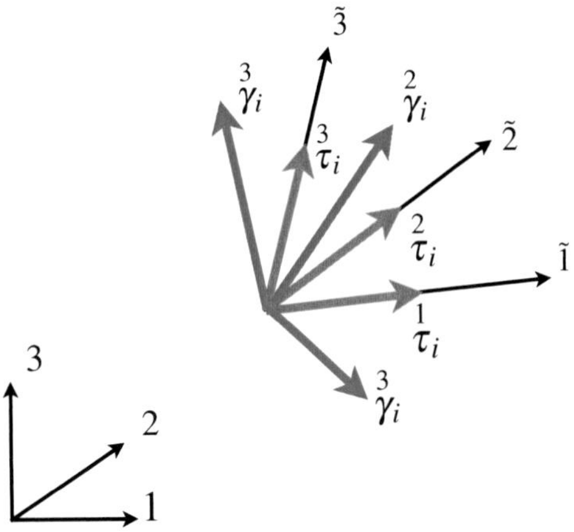

Figure 5.2: Non-orthogonal coordinates; the coordinates 1,2,3 are Cartesian; the coordinates $\tilde{1},\tilde{2},\tilde{3}$ are not. The three base vectors $\underset{m}{\tau}^i$ are tangent to the $\tilde{1},\tilde{2},\tilde{3}$ coordinates, the three gradient vectors $\overset{m}{\gamma}_i$ are orthogonal to the planes of the two coordinate axes that do not contain the number m.

and

$$\underset{m}{\tau}^i \overset{m}{\gamma}_j = \frac{\partial x_i}{\partial \tilde{x}_m}\frac{\partial \tilde{x}_m}{\partial x_j} = \delta_{ij}. \tag{5.34}$$

The dot product of tangential and gradient base vectors is zero because the tangential base vectors $\underset{m}{\tau}^i$ are normal to the gradient base vectors $\overset{n}{\gamma}_i$, except when $m = n$, in which case their dot product is one. The base vectors $\underset{m}{\tau}^i$ and $\overset{m}{\gamma}_i$ are generally not at right angles; the dot product of the base vectors is equal to

$$\underset{m}{\tau}^i \overset{n}{\gamma}_i = \underset{m}{\tau}\overset{n}{\gamma}\cos\overset{n}{\underset{m}{\alpha}} = \delta_{mn}, \tag{5.35}$$

where $\overset{n}{\underset{m}{\alpha}}$ is the angle between the base vectors. Since $\cos\overset{n}{\underset{m}{\alpha}} \neq 1$ unless $\overset{n}{\underset{m}{\alpha}} = \pi/2$, the lengths of the base vectors will generally not be equal to one, as in the Cartesian case. This is the result of the scale of measurement along the non-orthogonal coordinate axes being different from those along the Cartesian axes. In the present case the scales of measurement do not vary in space because the coordinate transformations are linear; the components of the base vectors are constants that depend upon the angles between the coordinate axes.

The components of a vector in the non-orthogonal system are not equal to their projections on the coordinate directions. There are two different sets of base vectors; we may decompose a vector in terms of components either along the tangent bases, or along the gradient bases. The former components are called the *contravariant* components and are denoted with the index in the place where normally an exponent appears, i.e., the contravariant components of a vector A_i in the transformed system are written as $\tilde{A}^i$. The Cartesian components of the vector A_i are related to the contravariant components $\tilde{A}^i$ as:

$$A_i = \underset{m}{\tau}^i \tilde{A}^m. \tag{5.36}$$

We may also decompose the vector in terms of components along the gradient bases. The corresponding components are the *covariant* components; we place the indices at the lower right corner of the symbol:

$$A_i = \overset{m}{\gamma}_i \tilde{A}_m \tag{5.37}$$

The two different kinds of decomposition into components are illustrated in Figure 5.3 for the two-dimensional case. The components $\tilde{A}_i$ and $\tilde{A}^i$ are not equal in length to the projections of A_i onto the base vectors; the projections consist of the dot product of $\tilde{A}^m$ and $\underset{m}{\tau}^i$ and that of $\tilde{A}_m$ and $\overset{m}{\gamma}_i$ and the base vectors are not of unit length. Since the covariant and contravariant components of the vectors are not equal to their projections, we distinguish them from their projections, called the physical components.

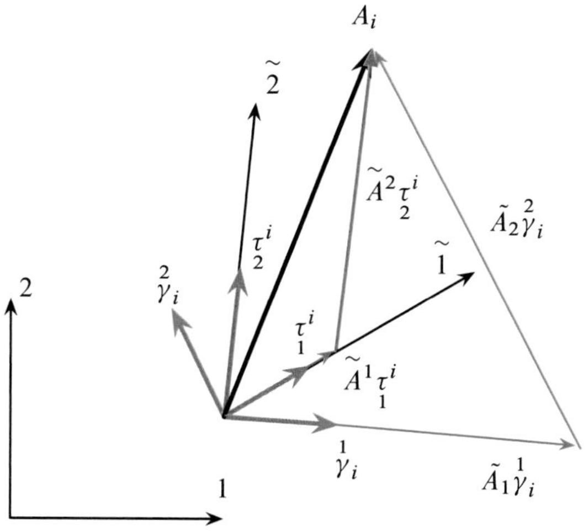

Figure 5.3: Decomposition of a vector A_i in its contra variant and covariant components

The Cartesian coordinate system is a special case, where covariant and contravariant components are equal; we place the index denoting Cartesian components at the lower right corner.

We determine a relationship between the contravariant and covariant components of the vector by first setting the right-hand sides of (5.36) and (5.37) equal to each other

$$\underset{m}{\tau}^i \tilde{A}^m = \overset{m}{\gamma}_i \tilde{A}_m, \tag{5.38}$$

and next contract both sides of this equation with $\overset{n}{\gamma}_i$:

$$\overset{n}{\gamma}_i \underset{m}{\tau}^i \tilde{A}^m = \overset{n}{\gamma}_i \overset{m}{\gamma}_i \tilde{A}_m. \tag{5.39}$$

We apply (5.34) to the left-hand side:

$$\delta_{nm}\tilde{A}^m = \tilde{A}^n = g^{nm}\tilde{A}_m. \tag{5.40}$$

The dot product of the covariant bases, g^{mn}, is called the *contravariant metric tensor*, and represents the scales of measurement in the directions normal to the axes. Indeed, the metric tensor is the dot product of the various covariant base vectors and thus indeed reflect the scales of measurement,

$$g^{nm} = \overset{n}{\gamma}_i \overset{m}{\gamma}_i. \tag{5.41}$$

If we contract both sides of (5.38) with $\underset{n}{\tau}^i$, we obtain:

$$\underset{m}{\tau}^i \underset{n}{\tau}^i \tilde{A}^m = \underset{n}{\tau}^i \overset{m}{\gamma}_i \tilde{A}_m = \delta_{nm}\tilde{A}_m = \tilde{A}_n, \tag{5.42}$$

or, introducing the *covariant metric tensor* g_{mn} as

$$g_{mn} = \underset{m}{\tau}^i \underset{n}{\tau}^i, \tag{5.43}$$

we obtain

$$\tilde{A}_n = g_{mn}\tilde{A}^m. \tag{5.44}$$

We say that contraction of the contravariant components $\tilde{A}^m$ with the covariant metric tensor g_{mn} 'brings the index down', i.e., results in producing the covariant components of the vector. Similarly, contraction of the covariant components $\tilde{A}_m$ with the contravariant metric tensor g^{mn} 'raises the index up', i.e., produces the contravariant components of the vector. We recall the definitions of the two metric tensors

$$g^{mn} = \overset{m}{\gamma}_i \overset{n}{\gamma}_i = \frac{\partial \tilde{x}_m}{\partial x_i}\frac{\partial \tilde{x}_n}{\partial x_i} \tag{5.45}$$

$$g_{mn} = \underset{m}{\tau}^i \underset{n}{\tau}^i = \frac{\partial x_i}{\partial \tilde{x}_m}\frac{\partial x_i}{\partial \tilde{x}_n}. \tag{5.46}$$

The two metric tensors are symmetrical.

We obtain the contravariant components $\tilde{A}^m$ from (5.36) by contraction with $\overset{n}{\gamma}_i$

$$A_i \overset{n}{\gamma}_i = \overset{n}{\gamma}_i \underset{m}{\tau}^i \tilde{A}^m = \tilde{A}^m \delta_{mn} = \tilde{A}^n, \tag{5.47}$$

and obtain the covariant components by contracting (5.37) with $\underset{n}{\tau}^i$

$$A_i \underset{n}{\tau}^i = \underset{n}{\tau}^i \overset{m}{\gamma}_i \tilde{A}_m = \delta_{mn}\tilde{A}_m = \tilde{A}_n. \tag{5.48}$$

The transformation rules for vectors are given by four equations in this case, two each for the contravariant and two each for the covariant components of the vector, i.e., by (5.36) with (5.47) and by (5.37) with (5.48)

Similar transformation rules apply to tensors; the rules for the contravariant components become

$$\tilde{A}^{mn} = \overset{m}{\gamma}_i \overset{n}{\gamma}_j A_{ij} \tag{5.49}$$

There exist a mix of components as well

$$\tilde{A}^{m}{}_{n} = \overset{m}{\gamma}_i \underset{n}{\tau}^j A_{ij}, \tag{5.50}$$

and

$$\tilde{A}_{m}{}^{n} = \underset{m}{\tau}^i \overset{n}{\gamma}_j A_{ij}. \tag{5.51}$$

5.3.1 Invariants

We concluded when dealing with Cartesian systems that contraction that eliminates all free indices results in a scalar quantity. This is not true for components in general coordinate systems. For example, the quantity A_{ii} is a scalar if the components of this tensor are in a Cartesian coordinate system. However, the quantities $\tilde{A}^{ii}$ and $\tilde{A}_{ii}$ are not invariants. We see this by expressing these quantities in terms of Cartesian coordinates

$$\tilde{A}^{mm} = \overset{m}{\gamma}_i \overset{m}{\gamma}_j A_{ij}, \tag{5.52}$$

which is not an invariant since the base vectors are unrelated to the components of the tensor A_{ij}. However, the quantities $\tilde{A}^{m}{}_{m}$ and $\tilde{A}_{m}{}^{m}$ are invariant scalars as follows from the transformation rules and (5.34),

$$\tilde{A}^{m}{}_{m} = \overset{m}{\gamma}_i \underset{m}{\tau}^j A_{ij} = \delta_{ij} A_{ij} = A_{ii}. \tag{5.53}$$

We obtain a similar result for $\tilde{A}_{m}{}^{m}$.

We create invariants by contracting indices in opposite positions, i.e., an index in the upper right position balanced by one in the lower right position of the contracting term. The dot product, an invariant, is

$$A_i A_i = \tilde{A}^i \tilde{A}_i, \tag{5.54}$$

which is invariant and matches $A_i A_i$ as we see by contracting (5.36) with (5.37),

$$A_i A_i = \underset{m}{\tau}^i \tilde{A}^m \overset{n}{\gamma}_i \tilde{A}_n = \delta_{mn} \tilde{A}^m \tilde{A}_n = \tilde{A}^m \tilde{A}_m. \tag{5.55}$$

5.3.2 The Gradient

Vectors are usually expressed in terms of their contravariant components, whereas the components of the gradient of a scalar are preferably expressed in terms of their covariant components. This becomes clear by considering the definition of the gradient of some scalar function F,

$$\frac{\partial F}{\partial \tilde{x}_i} = \frac{\partial F}{\partial x_m} \frac{\partial x_m}{\partial \tilde{x}_i} = \underset{i}{\tau}^m \frac{\partial F}{\partial x_m}, \tag{5.56}$$

which conforms to the transformation rule for a covariant vector, (5.48).

Of particular interest for later use is the transformation of the Laplacian, $L_{ii} = \partial_i \partial_i F$. The Laplacian is an invariant, so that we need to transform the gradient tensor L_{ij} as follows:

$$\tilde{L}^m{}_m = \overset{m}{\gamma}_i \underset{m}{\tau}^j L_{ij} = \delta_{ij} L_{ij} = L_{ii} = \nabla^2 F. \tag{5.57}$$

We discuss in the remainder of this chapter the stress, displacement gradient, and strain tensors. We do this in terms of Cartesian coordinates.

5.4 The Stress Tensor

The stress tensor is of the second rank, and operates on the vector normal to a plane to produce the stress vector, or traction, that operates on that plane,

$$t_i = \sigma_{ji} n_j, \tag{5.58}$$

where t_i represents the stress vector, and n_i the unit normal to the plane. The unit normals to the planes in the positive coordinate directions are:

$$\overset{m}{n}_i = \delta_{mi}, \tag{5.59}$$

where m represents the coordinate axis that serves as the normal to the plane considered. We apply (5.58) to the three normal vectors of (5.59) and obtain the following for the corresponding stress vectors $\overset{m}{t}_i$:

$$\overset{m}{t}_i = \sigma_{ji} \delta_{mj} = \sigma_{mi}. \tag{5.60}$$

It appears that the components σ_{mi} for $m = 1, 2, 3$ represent the three stress vectors, or tractions, on the planes with normals in the $1, 2, 3$ directions, respectively, see Figure 5.4. The stresses that act on the planes with their normals in the negative coordinate directions are obtained from (5.58) by replacing n_i by $\overset{m}{n}_i$ with

$$\overset{m}{n}_i = -\delta_{mi}, \tag{5.61}$$

so that (5.58) becomes

$$\overset{m}{t}_i = -\sigma_{ji} \delta_{mi} = -\sigma_{mi}. \tag{5.62}$$

The expression σ_{ii} is a scalar since there is no free index, and is invariant to a coordinate transformation; this invariant is called the first invariant of the stress tensor. The isotropic stress is defined as one-third of the first invariant and is usually represented as $\overset{0}{\sigma}$

$$\overset{0}{\sigma} = \tfrac{1}{3} \sigma_{ii}. \tag{5.63}$$

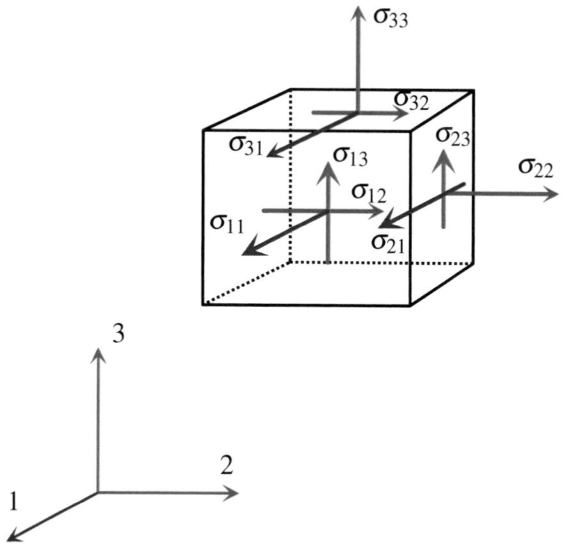

Figure 5.4: Definition of the components of the stress tensor.

5.4.1 The Normal and Shear Stress Components

We decompose the stress vector that acts on a plane into two components: the normal stress and the shear stress; the former is normal to the plane, and the latter acts in the plane. We contract t_i with n_i to obtain the normal stress:

$$\overset{n}{t} = t_i n_i = \sigma_{ij} n_i n_j. \tag{5.64}$$

The normal stress is positive if pointing in the direction of the unit normal n_i. We multiply $\overset{n}{t}$ with n_j to obtain the normal stress as a vector:

$$\overset{n}{t}_j = t_i n_i n_j = \sigma_{nm} n_n n_m n_j. \tag{5.65}$$

The tangential component of the stress vector is the difference between the stress vector t_i and its normal component:

$$\overset{s}{t}_j = t_j - t_i n_i n_j. \tag{5.66}$$

5.4.2 Principal Directions and Principal Values of the Stress Tensor in Two Dimensions.

We determine the principal directions and the principal values of the stress tensor in two dimensions for the case of plane stress, i.e., the case that there are no stresses in planes with normals in the 3-direction.

Principal Directions

The principal directions are the directions of the normals to the planes that carry only normal stresses, i.e., the planes where the shear stresses are zero. We rotate the 1,2-coordinate system in such a way that the shear stress on the plane with its normal in the $\tilde{1}$-direction is zero.

The transformed stress tensor $\tilde{\sigma}_{1j} = a_{m1}a_{nj}\sigma_{mn}$ represents the stresses on the plane normal to the $\tilde{1}$ axis and $\tilde{\sigma}_{12} = 0$,

$$\tilde{\sigma}_{12} = 0 = a_{m1}a_{n2}\sigma_{mn}, \tag{5.67}$$

where

$$a_{mj} = \underset{j}{\tau}^{m}. \tag{5.68}$$

If we represent the angle between the 1-axis and the $\tilde{1}$-axis as ψ then

$$\begin{aligned} a_{m1} &= \underset{1}{\tau}^{m} = (\ \ \cos\psi, \sin\psi) \\ a_{m2} &= \underset{2}{\tau}^{m} = (-\sin\psi, \cos\psi). \end{aligned} \tag{5.69}$$

We work out (5.67):

$$\begin{aligned} 0 &= a_{11}a_{12}\sigma_{11} + a_{11}a_{22}\sigma_{12} + a_{21}a_{12}\sigma_{21} + a_{21}a_{22}\sigma_{22} \\ &= -\sigma_{11}\sin\psi\cos\psi + \sigma_{12}\cos^2\psi - \sigma_{12}\sin^2\psi + \sigma_{22}\sin\psi\cos\psi \\ &= -\tfrac{1}{2}\sin(2\psi)(\sigma_{11} - \sigma_{22}) + \sigma_{12}\cos(2\psi). \end{aligned} \tag{5.70}$$

We solve for the tangent of 2ψ, i.e., the tangent of twice the angle that the principal direction makes with the 1-axis,

$$\tan(2\psi) = \frac{\sigma_{12}}{\frac{1}{2}(\sigma_{11} - \sigma_{22})}. \tag{5.71}$$

Principal Values

With the principal directions known, we determine the principal values in terms of the principal directions. We obtain an expression for the principal stress σ_{11} from $\tilde{\sigma}_{ij} = a_{mi}a_{nj}\sigma_{mn}$ by setting both i and j equal to 1:

$$\begin{aligned} \tilde{\sigma}_{11} &= a_{m1}a_{n1}\sigma_{mn} = a_{11}^2\sigma_{11} + a_{21}a_{11}\sigma_{21} + a_{11}a_{21}\sigma_{12} + a_{21}^2\sigma_{22} \\ &= \sigma_{11}\cos^2\psi + 2\sigma_{12}\sin\psi\cos\psi + \sigma_{22}\sin^2\psi \\ &= \tfrac{1}{2}\sigma_{11}[1 + \cos(2\psi)] + \tfrac{1}{2}\sigma_{22}[1 - \cos(2\psi)] + \sigma_{12}\sin(2\psi) \\ &= \tfrac{1}{2}(\sigma_{11} + \sigma_{22}) + \tfrac{1}{2}(\sigma_{11} - \sigma_{22})\cos(2\psi) + \sigma_{12}\sin(2\psi). \end{aligned} \tag{5.72}$$

The second principal stress acts on a plane that is at right angles to the plane with its normal in the direction ψ. Since the angle ψ is multiplied by a factor 2, and since the signs of both the cosine and the sine function change upon adding π to the argument, we obtain for $\tilde{\sigma}_{22}$:

$$\tilde{\sigma}_{22} = \tfrac{1}{2}(\sigma_{11} + \sigma_{22}) - \tfrac{1}{2}(\sigma_{11} - \sigma_{22})\cos(2\psi) - \sigma_{12}\sin(2\psi). \tag{5.73}$$

5.4.3 Stress Deviator

The difference between the major and minor principal stresses is the *stress deviator*, half of which we represent by the symbol λ. It is sometimes useful to express the three components of the stress tensor in terms of λ, ψ, and $\overset{0}{\sigma}$. We write the major principal stress, (5.72), as $\overset{0}{\sigma}+\lambda$, and subtract $\overset{0}{\sigma}$ from both sides of the equation,

$$\lambda = \tfrac{1}{2}(\sigma_{11}-\sigma_{22})\cos(2\psi)+\sigma_{12}\sin(2\psi). \tag{5.74}$$

We use (5.71),

$$\tfrac{1}{2}(\sigma_{11}-\sigma_{22})\sin(2\psi)-\sigma_{12}\cos(2\psi)=0, \tag{5.75}$$

multiply (5.74) by $\cos(2\psi)$ and (5.75) by $\sin(2\psi)$, and add:

$$\lambda\cos(2\psi)=\tfrac{1}{2}(\sigma_{11}-\sigma_{22}). \tag{5.76}$$

We find the expression for σ_{12} by multiplying (5.74) by $\sin(2\psi)$, multiplying (5.75) by $\cos(2\psi)$, and subtracting the results,

$$\lambda\sin(2\psi)=\sigma_{12}. \tag{5.77}$$

We first square, then add, the two sides of the latter two equations in order to obtain an expression for λ in terms of the components of the stress tensor:

$$\lambda^2=\left[\tfrac{1}{2}(\sigma_{11}-\sigma_{22})\right]^2+\sigma_{12}^2. \tag{5.78}$$

We find expressions for σ_{11} and σ_{22} from (5.76) and $\overset{0}{\sigma}=\frac{1}{2}(\sigma_{11}+\sigma_{22})$,

$$\begin{aligned}\sigma_{11}&=\overset{0}{\sigma}+\lambda\cos(2\psi)\\ \sigma_{22}&=\overset{0}{\sigma}-\lambda\cos(2\psi).\end{aligned} \tag{5.79}$$

5.4.4 Invariants

The first invariant of the stress tensor, σ_{ii}, equals twice the isotropic stress,

$$\underset{1}{I}=\sigma_{kk}=2\overset{0}{\sigma}. \tag{5.80}$$

We obtain the expression for a second invariant by contracting σ_{ij} with itself:

$$\underset{2}{I}=\sigma_{ij}\sigma_{ij}. \tag{5.81}$$

The stress deviator is also an invariant. In order to demonstrate this, we form the combination:

$$\begin{aligned}\underset{2}{I}-\underset{1}{I}^2&=\sigma_{ij}\sigma_{ij}-\sigma_{kk}\sigma_{mm}\\ &=\sigma_{11}^2+2\sigma_{12}^2+\sigma_{22}^2-(\sigma_{11}+\sigma_{22})^2=2\sigma_{12}^2-2\sigma_{11}\sigma_{22}.\end{aligned} \tag{5.82}$$

We expand the expression for λ^2, (5.78),

$$\lambda^2 = \tfrac{1}{4}\left[\sigma_{11}^2 + \sigma_{22}^2 - 2\sigma_{11}\sigma_{22}\right] + \sigma_{12}^2 = \tfrac{1}{4}\left[\sigma_{11}^2 + \sigma_{22}^2 + 2\sigma_{11}\sigma_{22}\right] - \sigma_{11}\sigma_{22} + \sigma_{12}^2, \tag{5.83}$$

or with (5.82),

$$\lambda^2 = \tfrac{1}{4}\sigma_{kk}^2 + \tfrac{1}{2}(-I_1^2 + I_2) = -\tfrac{1}{4}I_1^2 + \tfrac{1}{2}I_2, \tag{5.84}$$

which is indeed an invariant.

Problem 5.2

Derive equations for the angle ψ between the major principal stress and the 1-axis, using (5.66), by setting the shear stress component equal to zero, expressing the traction in terms of the stress tensor and the unit normal to the plane. Compare your answer with (5.71).

5.4.5 The Equilibrium Conditions

The stresses that act on the six planes of the cube in Figure 5.4 act on opposite planes in opposite direction from one another. Thus, if the stresses are constant, they are in equilibrium. The situation changes if the stresses vary with position. We apply the condition of equilibrium to the stresses in the 2-direction on a volume of sides $\Delta x_1, \Delta x_2, \Delta x_3$ and centered at x_i, see Figure 5.5

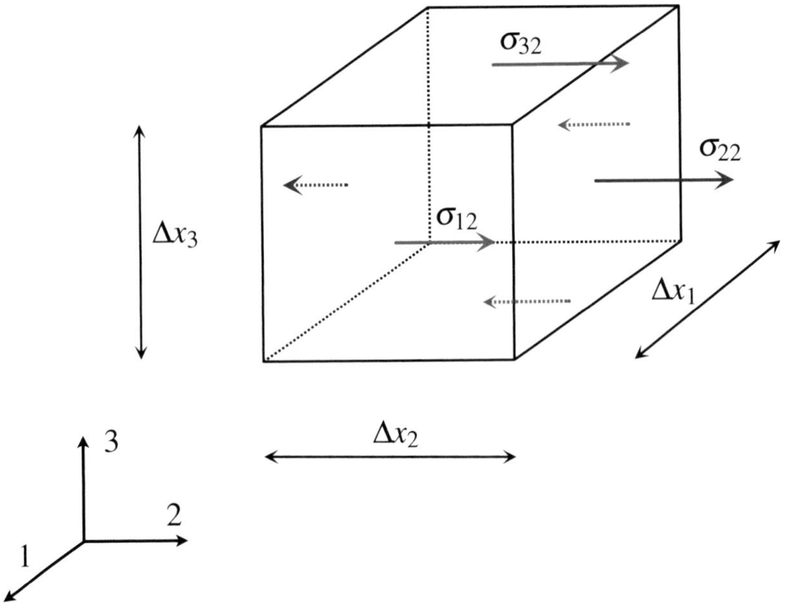

Figure 5.5: Equilibrium in the 2-direction

$$\begin{aligned}&[\sigma_{22}(x_1,x_2+\tfrac{1}{2}\Delta x_2,x_3)-\sigma_{22}(x_1,x_2-\tfrac{1}{2}\Delta x_2,x_3)]\Delta x_1\Delta x_3\\&+[\sigma_{12}(x_1+\tfrac{1}{2}\Delta x_1,x_2,x_3)-\sigma_{12}(x_1-\tfrac{1}{2}\Delta x_1,x_2,x_3)]\Delta x_2\Delta x_3\\&+[\sigma_{32}(x_1,x_2,x_3+\tfrac{1}{2}\Delta x_3)-\sigma_{32}(x_1,x_2,x_3-\tfrac{1}{2}\Delta x_3)]\Delta x_1\Delta x_2\\&+B_2\Delta x_1\Delta x_2\Delta x_3=F_2\Delta x_1\Delta x_2\Delta x_3.\end{aligned}\tag{5.85}$$

where B_2 is the component of the body force B_i per unit volume. The force F_i, also per unit volume, is the resultant force due to the stresses and the body force B_i. This force is zero if the system is in equilibrium; if it is not, then $F_i = \rho e_i$, where ρ is the mass per unit volume, i.e., the density, and e_i is a unit vector pointing in the direction of the resultant force. We pass to the limit for $\Delta x_1 \to 0, \Delta x_2 \to 0, \Delta x_3 \to 0$ and obtain

$$\partial_i \sigma_{i2} + B_2 = F_2 = \rho a_2. \tag{5.86}$$

We generalize this equation to apply to all three coordinate directions,

$$\partial_i \sigma_{ij} + B_j = F_j = \rho a_j. \tag{5.87}$$

5.5 The Displacement Gradient and Strain Tensors

We consider a deforming body and represent the displacements of points in that body relative to some fixed initial state as u_i. We begin by examining a linear displacement, and consider displacement vectors of two points, $\overset{\mathscr{P}}{u}_i$ at $\mathscr{P}$ and $\overset{\mathscr{Q}}{u}_i$ at $\mathscr{Q}$. The vector $l\nu_i$, where ν_i is a unit vector, points from point $\mathscr{P}$ to point $\mathscr{Q}$

$$l\nu_i = \overset{\mathscr{Q}}{x}_i - \overset{\mathscr{P}}{x}_i, \tag{5.88}$$

where $\overset{\mathscr{P}}{x}_i$ and $\overset{\mathscr{Q}}{x}_i$ are the coordinates of points $\mathscr{P}$ and $\mathscr{Q}$, see Figure 5.6. The relative displacement per unit length is a_i, with

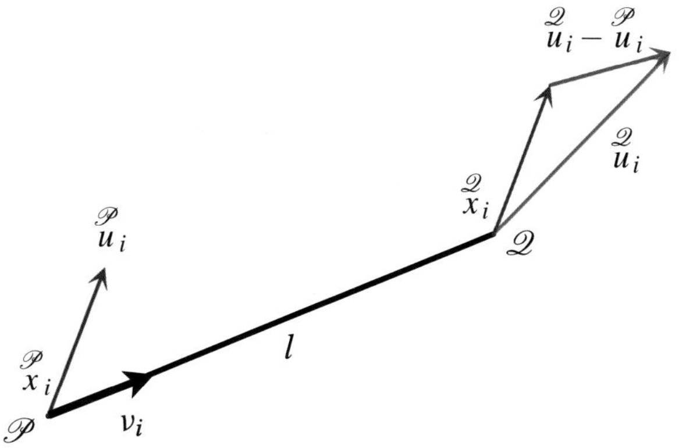

Figure 5.6: Relative displacement of the endpoints of a filament of length l with orientation ν_i

$$a_i = \frac{\Delta u_i}{l} = \frac{\overset{\mathcal{Q}}{u}_i - \overset{\mathcal{P}}{u}_i}{l} \tag{5.89}$$

The relative displacement vector a_i is useful to describe the deformation of a body that experiences linear displacements.

5.5.1 The Displacement Gradient Tensor

The displacements generally vary with position; we use the displacement gradient tensor rather than the relative displacement vector to describe the deformation of the body:

$$\alpha_{ij} = \partial_i u_j = \frac{\partial u_j}{\partial x_i}. \tag{5.90}$$

We express the relative displacement vector a_i in terms of the displacement gradient tensor as:

$$a_j = \lim_{|\Delta x_i| \to 0} \frac{\partial u_j}{\partial x_i} \frac{\Delta x_i}{\sqrt{\Delta x_m \Delta x_m}} = \alpha_{ij} v_i, \tag{5.91}$$

where v_i is a unit vector in the direction of the filament considered. This equation is similar to that obtained for the stress vector acting on an arbitrary plane.

The vector a_i represents the relative displacement of the end points of a filament of incremental length oriented in the direction v_i. As for stresses, we define relative displacements in the direction normal to v_i and in the direction of v_i. The latter displacement represents the stretch of the filament, whereas the former represents rotation of the filament. The relative rotation of two filaments that are originally at right angles represents a change in angle between the filaments, the shear strain γ. The rotation of both filaments together represents local body rotation. We distinguish between the relative and absolute rotations of the filaments by separating the symmetrical and anti-symmetrical parts of the displacement gradient tensor:

$$\partial_i u_j = e_{ij} + \omega_{ij}. \tag{5.92}$$

5.5.2 The Strain Tensor and the Rotation Tensor

The strain tensor e_{ij} is the symmetrical part of the displacement gradient tensor:

$$e_{ij} = \tfrac{1}{2}(\partial_i u_j + \partial_j u_i). \tag{5.93}$$

We illustrate application of the strain tensor by considering strain of two filaments at ninety degrees from one another, see Figure 5.7; the first filament has an orientation $v_i = \delta_{1i}$ and the second one has an orientation $v_i = \delta_{2i}$. The relative displacements e_i for these filaments are:

$$\begin{aligned} a_j &= e_{ij} v_i = e_{ij} \delta_{1i} = e_{1j} \\ a_j &= e_{ij} v_i = e_{ij} \delta_{2i} = e_{2j} \end{aligned} \tag{5.94}$$

These strains are shown in Figure 5.7; the strains e_{11} and e_{22} represent stretching and the strains $e_{12} = e_{21}$ represent a change in angle between the two filaments; the ends of the filaments move toward one another over a distance $e_{12} = e_{21}$.

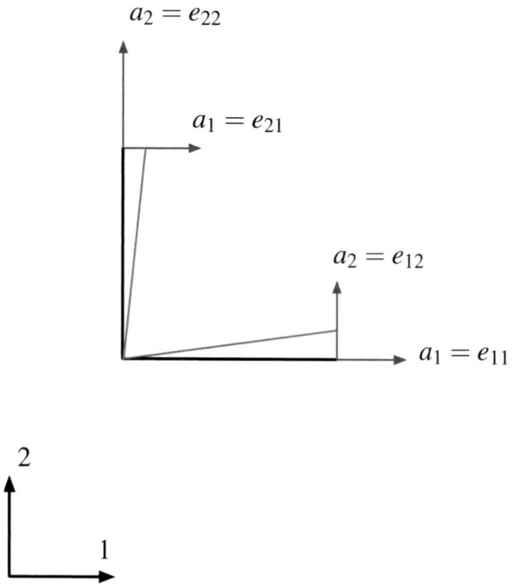

Figure 5.7: Strains in the 1,2 plane

The rotation tensor ω_{ij} represents the anti-symmetrical part of the displacement gradient tensor:

$$\omega_{ij} = \tfrac{1}{2}(\partial_i u_j - \partial_j u_i). \tag{5.95}$$

We use the relationship between the epsilon and delta tensors, (3.75) to write

$$\omega_{ij} = \tfrac{1}{2}(\partial_i u_j - \partial_j u_i) = \tfrac{1}{2}(\delta_{ip}\delta_{jq} - \delta_{iq}\delta_{jp})\partial_p u_q = \tfrac{1}{2}\varepsilon_{mij}\varepsilon_{mpq}\partial_p u_q = \tfrac{1}{2}\varepsilon_{mij}C_m, \tag{5.96}$$

where C_m is the curl of the displacement vector. It follows that the rotation tensor is fully defined by three components in three dimensions.

For the two-dimensional case, i and j range from 1 to 2 and the only non-zero component in (5.96) exists for $m = 3$,

$$\omega_{ij} = \tfrac{1}{2}\varepsilon_{3ij}C_3. \tag{5.97}$$

The rotation tensor in two dimensions is fully defined by a single scalar, ω, where $\omega = \frac{1}{2}C_3$, so that (5.96) reduces to

$$\omega_{ij} = \omega\varepsilon_{3ij}. \tag{5.98}$$

We demonstrate the effect of the rotation tensor ω_{ij} as for the strain tensor, by considering two orthogonal filaments. We obtain the following for the displacement vector:

$$\begin{aligned} v_i &= \delta_{1i} \qquad a_j = \omega_{ij}v_i = \omega_{ij}\delta_{1i} = \omega_{1j} = \tfrac{1}{2}\varepsilon_{31j}\omega_3 = (0, \omega) \\ v_i &= \delta_{2i} \qquad a_j = \omega_{ij}v_i = \omega_{ij}\delta_{2i} = \omega_{2j} = \tfrac{1}{2}\varepsilon_{32j}\omega_3 = (-\omega, 0), \end{aligned} \tag{5.99}$$

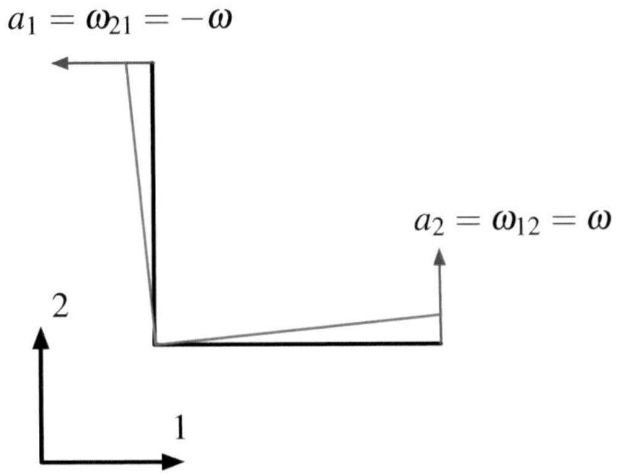

Figure 5.8: Rotation in the 1,2 plane.

or:

$$\begin{aligned}
\omega_{ij} &= \tfrac{1}{2}\varepsilon_{1ij}C_1 + \tfrac{1}{2}\varepsilon_{2ij}C_2 + \tfrac{1}{2}\varepsilon_{3ij}C_3 \\
\omega_{12} &= \tfrac{1}{2}\varepsilon_{312}C_3 = \tfrac{1}{2}C_3 = \omega \\
\omega_{21} &= \tfrac{1}{2}\varepsilon_{321}C_3 = -\tfrac{1}{2}C_3 = -\omega,
\end{aligned} \tag{5.100}$$

The rotation tensor indeed represents true rotation, but only in an incremental sense; this is true only for small displacement gradients. That the rotation tensor represents pure rotation is in accordance with (5.96); the displacement vector is:

$$a_j = \omega_{ij} v_i = \tfrac{1}{2}\varepsilon_{mij} v_i C_m. \tag{5.101}$$

The vector a_i is the cross product of the vector v_i, which points in the direction of the filament, and the curl of the displacement vector; for the two-dimensional case, illustrated in Figure 5.8, the curl points in the 3-direction, the displacement is normal to v_i, and is always oriented the same way with respect to v_i.

As for the stress tensor, the first invariant of the strain tensor plays a special role in stress-strain relationships, for example in the theory of linear elasticity, discussed in Section 9.5. The first invariant e_{ii} is the volume strain

$$e_{ii} = \partial_i u_i, \tag{5.102}$$

which is the divergence of the displacement vector.

5.6 Integrability; The Compatibility Conditions

We consider a vector or tensor that is defined as the gradient of a scalar or a tensor of lower rank. It is generally not true that integration of the higher order tensor yields the lower order scalar or tensor, from which it was derived. This follows from the notion that the derived quantity has a greater number of components than the original one;

degrees of freedom have been added and these degrees of freedom must be subjected to constraints in order to make them compatible with the fewer number of degrees of freedom of the original quantity.

5.6.1 The Gradient of a Scalar

The conditions of integrability of a vector, defined as the gradient of a scalar,

$$v_i = -\partial_i \Phi, \tag{5.103}$$

are that the vector field is irrotational:

$$C_i = \varepsilon_{ijk}\partial_j v_k = \varepsilon_{ijk}\partial_j\partial_k\Phi = 0. \tag{5.104}$$

The three homogeneous equations (5.104), which the three components of v_i must satisfy, are sufficient to guarantee that equations (5.103) are integrable.

5.6.2 The Gradient of a Vector

A similar case applies if a tensor, for example the strain tensor e_{ij}, is derived from the gradient of a vector field. We obtain the integrability conditions as for the case of a vector. We first consider the definition of the displacement gradient tensor, in terms of the displacement vector u_i,

$$\alpha_{km} = \partial_k u_m. \tag{5.105}$$

Integrability of these equations requires that

$$\varepsilon_{ijk}\partial_j\alpha_{km} = \varepsilon_{ijk}\partial_j\partial_k u_m = 0. \tag{5.106}$$

We obtain the integrability conditions for the strains by using the definitions of the strain and rotation tensors:

$$\varepsilon_{ijk}\partial_j\alpha_{km} = \varepsilon_{ijk}\partial_j e_{km} + \varepsilon_{ijk}\partial_j\omega_{km}. \tag{5.107}$$

We use (5.96) to rewrite the trailing term as

$$\varepsilon_{ijk}\partial_j\omega_{km} = \tfrac{1}{2}\varepsilon_{ijk}\varepsilon_{kmn}\partial_j C_n = \tfrac{1}{2}(\delta_{im}\delta_{jn} - \delta_{in}\delta_{jm})\partial_j C_n = \tfrac{1}{2}\delta_{im}\partial_j C_j - \tfrac{1}{2}\partial_m C_i, \tag{5.108}$$

where C_n is the curl,

$$C_n = \varepsilon_{npq}\partial_p u_q. \tag{5.109}$$

The curl is divergence-free, as is seen from (5.109) by taking the divergence, so that (5.108) reduces to

$$\varepsilon_{ijk}\partial_j\omega_{km} = -\tfrac{1}{2}\partial_m C_i. \tag{5.110}$$

We combine (5.107) with (5.110),

$$\varepsilon_{ijk}\partial_j\alpha_{km} = \varepsilon_{ijk}\partial_j e_{km} - \tfrac{1}{2}\partial_m C_i. \tag{5.111}$$

These equations represent nine conditions for the six independent components of the strain tensor and the three components of the curl; we eliminate the curl in what follows.

Equation (5.111) represents a tensor with free indices i and m, which we call A_{im} and take the curl, $\varepsilon_{pnm}\partial_n A_{im}$, which gives:

$$\varepsilon_{pnm}\partial_n A_{im} = \varepsilon_{pnm}\varepsilon_{ijk}\partial_j\partial_n\alpha_{km} = \varepsilon_{ijk}\varepsilon_{pnm}\partial_j\partial_n\partial_k u_m = 0, \tag{5.112}$$

because $\partial_j\partial_k = \partial_k\partial_j$. We replace A_{im} by (5.111),

$$\varepsilon_{pnm}\partial A_{im} = \varepsilon_{pnm}\varepsilon_{ijk}\partial_n\partial_j\varepsilon_{km} - \tfrac{1}{2}\varepsilon_{pnm}\partial_m\partial_n C_i = 0, \tag{5.113}$$

so that:

$$\varepsilon_{pnm}\varepsilon_{ijk}\partial_n\partial_j e_{km} = \tfrac{1}{2}\varepsilon_{pnm}\partial_m\partial_n C_i = 0. \tag{5.114}$$

These equations are the nine integrability conditions that must be imposed on the components of the strain tensor. If these conditions are met, integration of the strains produces a set of unique displacements. Three of the nine equations occur twice as a result of the symmetry of both the operators $\partial_i\partial_j$ and the strain tensor e_{ij}. We demonstrate that equations (5.114) are symmetrical in terms of the indices p and i by contracting with ε_{piq} and showing that the result is zero:

$$\begin{aligned}\varepsilon_{piq}\varepsilon_{pmn}\varepsilon_{ijk}\partial_n\partial_j e_{km} &= (\delta_{im}\delta_{qn} - \delta_{in}\delta_{qm})\varepsilon_{ijk}\partial_n\partial_j e_{km}\\ &= \varepsilon_{mjk}\partial_q\partial_j e_{km} - \varepsilon_{ijk}\partial_i\partial_j e_{kq}\end{aligned} \tag{5.115}$$

This represents three equations, one for each of the three possible values of q. The first term is zero because of the symmetry of e_{km} and the second term is zero because of the symmetry of $\partial_i\partial_j$, so that

$$\varepsilon_{piq}\varepsilon_{pmn}\varepsilon_{ijk}\partial_n\partial_j e_{km} = 0. \tag{5.116}$$

The equations obtained from (5.114) are thus indeed symmetrical in terms of the indices i and p. The nine equations (5.114) for the strain tensor components,

$$\varepsilon_{pmn}\varepsilon_{ijk}\partial_n\partial_j e_{km} = 0, \tag{5.117}$$

reduce to the six independent ones, obtained by letting p and i assume the following values:

$$\begin{aligned} &p = 1 && i = 1,2,3\\ &p = 2 && i = 2,3\\ &p = 3 && i = 3. \end{aligned} \tag{5.118}$$

These equations are the compatibility conditions for the strain tensor for small displacement gradients.

Problem 5.3

Work out the compatibility equations for plane strain (i.e., for the case that $e_{m3} = 0, m = 1,2,3$.

Problem 5.4

Work out the six compatibility equations in terms of $(x,y,z) \equiv x_i$ and compare your result with the equations as found in the literature.

5.7 The Velocity Gradient Tensor

The velocity gradient tensor plays a role in fluids similar to that of the displacement gradient tensor for solids. The velocity gradient tensor is:

$$\dot{\alpha}_{ij} = \partial_i v_j, \tag{5.119}$$

where v_j is the velocity. As for the displacements, we separate the velocity gradient tensor into its symmetrical and anti-symmetrical parts:

$$\dot{\alpha}_{ij} = \tfrac{1}{2}(\dot{\alpha}_{ij} + \dot{\alpha}_{ji}) + \tfrac{1}{2}(\dot{\alpha}_{ij} - \dot{\alpha}_{ji}). \tag{5.120}$$

The symmetrical part is $\dot{e}_{ij}$:

$$\dot{e}_{ij} = \tfrac{1}{2}(\partial_i v_j + \partial_j v_i). \tag{5.121}$$

The anti-symmetrical part is (compare (5.96))

$$\dot{\omega}_{ij} = \tfrac{1}{2}(\partial_i v_j - \partial_j v_i) = \varepsilon_{ijm} C_m, \tag{5.122}$$

where C_m is the curl. The divergence of the velocity field plays a role similar to that of the volume strain for the displacement gradient and is sometimes represented as D

$$D = \partial_i v_i. \tag{5.123}$$

Chapter 6

Partial Differential Equations of the First Order

We consider linear partial differential equations of the first order. A partial differential equation is linear if the dependent variables, i.e., the functions that are being solved for, appear only to the first power and the coefficients, i.e., the factors the derivatives are multiplied by, do not contain the dependent variables. Solutions to linear partial differential equations can be combined to form new ones; superposition of solutions produces solutions that satisfy the differential equation.

6.1 The General Differential Equation of the First Order

We write the general first-order partial differential equation in three dimensions in the form

$$A_1\frac{\partial u}{\partial x}+A_2\frac{\partial u}{\partial y}+A_3\frac{\partial u}{\partial z}=E. \tag{6.1}$$

We write this in indicial notation:

$$A_i\frac{\partial u}{\partial x_i}=E. \tag{6.2}$$

The left hand side of this equation represents the projection of the gradient of u onto the direction of A_i, as illustrated in Figure 6.1. Let s represent the arc length along a curve tangent to A_i, i.e.,

$$\frac{dx_i}{ds}=\frac{A_i}{A}, \tag{6.3}$$

where we limit the modulus A to be unequal to zero,

$$A=\sqrt{A_iA_i}\neq 0. \tag{6.4}$$

We write the projection of $\partial u/\partial x_i$ onto A_i as du/ds; the dot product at the left in (6.2)

O. D. L. Strack, *Applications of Vector Analysis and Complex Variables in Engineering*, https://doi.org/10.1007/978-3-030-41168-8_6

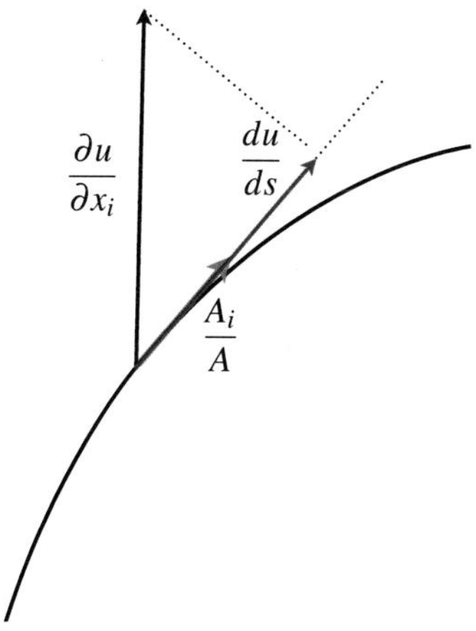

Figure 6.1: Projection of $\partial_i u$ onto the characteristic

is equal to du/ds times A. We use that $A \neq 0$ and divide both sides of the differential equation by A,

$$\frac{A_i}{A}\partial_i u = \frac{du}{ds} = \frac{E}{A}. \tag{6.5}$$

We replaced the original partial differential equation by a single ordinary differential equation written along a curve s, plus equations, (6.3), that define a set of curves with special meaning for this partial differential equation; they are its *characteristics.*

We see from (6.5) that the differential equation only gives information along the characteristics; discontinuities in both u and the derivative of u normal to a characteristic can occur along it without violating the partial differential equation. Figure 6.2 is an illustration of this property for the special case that u is a function of x_1 and x_2 only. We represent the function $u = u(x_1, x_2)$ as a surface above the $(1,2)$-plane. The differential equation (6.5) can be integrated along the characteristics, provided that the boundary values of u are given at one, and only one, point of each characteristic. Because the computation occurs independently for each characteristic (information cannot travel across the characteristics) discontinuities in boundary values of u will propagate along these curves.

A boundary value problem for a first-order partial differential equation is well-posed i.e., solvable, only if boundary values are applied along a boundary that intersects the characteristics only once; boundary values cannot be applied along a characteristic. Boundary values must be known on either boundary $\mathscr{A}$ or on $\mathscr{B}$ in Figure 6.2, but not on both.

6.1.1 Advective Contaminant Transport

We consider advective contaminant transport in a fluid as an example of a first-order partial differential equation. Advective transport implies that the contaminant is carried by the fluid; the effect of diffusion or dispersion is not considered. The mass flux is cv_i,

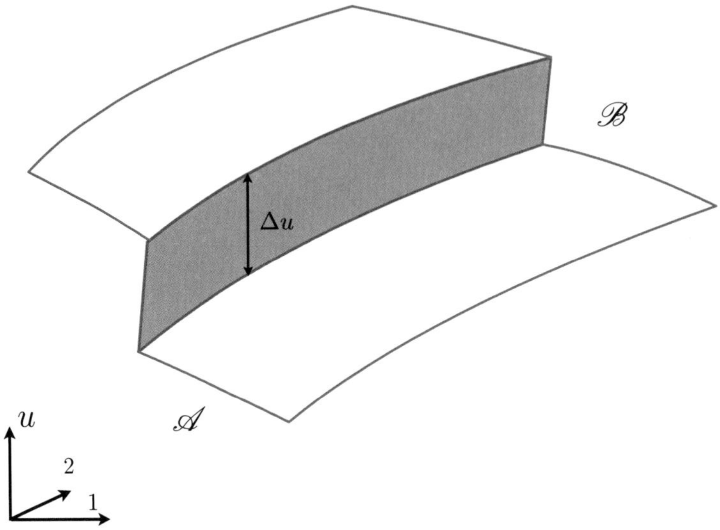

Figure 6.2: Discontinuity along characteristics

where c is the concentration [M/L^3] and v_i the velocity of the fluid. The mass-balance equation is:

$$\frac{\partial (cv_i)}{\partial x_i} = v_i \frac{\partial c}{\partial x_i} + c \frac{\partial v_i}{\partial x_i} = -\lambda c - \frac{\partial c}{\partial t}, \tag{6.6}$$

where the term $-\lambda c$ represents the loss of mass due to decay, and λ [1/T] is the decay rate factor. The divergence of the velocity is zero, i.e., $\partial_i v_i = 0$, if the fluid is incompressible; in that case:

$$v_i \frac{\partial c}{\partial x_i} = -\lambda c - \frac{\partial c}{\partial t}, \tag{6.7}$$

or

$$\frac{\partial c}{\partial t} + v_i \frac{\partial c}{\partial x_i} = -\lambda c. \tag{6.8}$$

The term $v_i \partial_i c$ represents the dot product of the velocity vector and the gradient of the concentration, as shown in Figure 6.3. The projection of $\partial_i c$ onto v_i is the directional derivative dc/ds, with s the arc length measured along a curve tangent to v_i, so that the second term of (6.8) can be written as $v dc/ds$,

$$\frac{\partial c}{\partial t} + v \frac{dc}{ds} = -\lambda c, \tag{6.9}$$

where v is the magnitude of the velocity vector. If s is the distance traveled of a point moving along the path line with velocity v, then $v = ds/dt$, and (6.9) becomes

$$\frac{\partial c}{\partial t} + \frac{dc}{ds}\frac{ds}{dt} = -\lambda c. \tag{6.10}$$

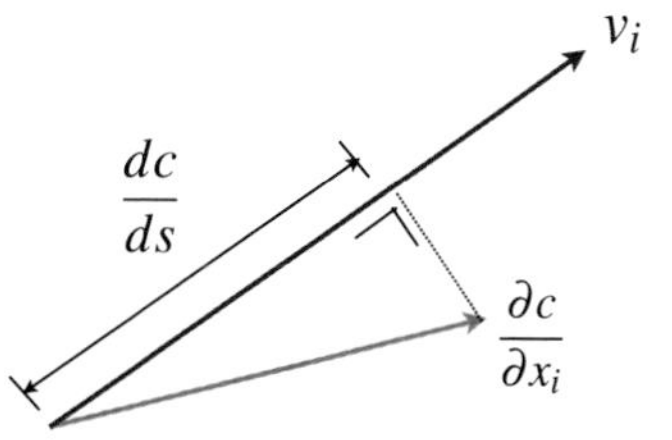

Figure 6.3: The dot product $v_i \frac{\partial c}{\partial x_i} = v \frac{dc}{ds}$

The left-hand side of this equation represents the material-time derivative,

$$\frac{Dc}{Dt} = -\lambda c \tag{6.11}$$

This equation represents the partial differential equation written along the characteristic, which in this case involves both space and time. The solution of this ordinary differential equation is

$$c = c_0 e^{-\lambda(t-t_0)}, \tag{6.12}$$

where c_0 is the concentration at time t_0, the starting time on the path line. We must recognize that (6.11) represents the concentration at a point, moving with velocity v along a path line. Thus, neither (6.11), nor (6.12) is the complete solution to the problem; we must know *where* (6.12) applies. We obtain corresponding values of s and t from

$$v = \frac{ds}{dt}, \tag{6.13}$$

or

$$t - t_0 = \int_{s_0}^{s} \frac{ds}{v} = f(s). \tag{6.14}$$

We show the functional relationship (6.14) graphically in the s,t-plane in Figure 6.4. Different curves exist for different values of t_0 and s_0. These curves are called the characteristics of the differential equation. Recall that discontinuities can occur along characteristics in both the function and the derivative normal to the characteristic.

We apply (6.11) to solve the problem of movement of a contaminant, without decay, i.e., $\lambda = 0$, originally occupying a domain $\mathscr{D}$ in the flow field, as shown in Figure 6.5. The concentration inside $\mathscr{D}$ at time $t = t_0$ is c_0. Then, according to the solution given by (6.12) and (6.14), the contaminated domain $\mathscr{D}$ will move through the flow field, thereby changing its shape to a domain $\mathscr{D}'$ at time t (see Figure 6.5). The concentration at t will be zero everywhere outside $\mathscr{D}'$, and equal to c_0 inside $\mathscr{D}'$. We see that the initial discontinuity in concentration travels along the characteristics with the flowing fluid.

In reality there are other processes such as dispersion and diffusion, causing the sharp edges of the contaminant plume to become smooth. In some applications the plug flow associated with purely convective transport is a good first-order approximation.

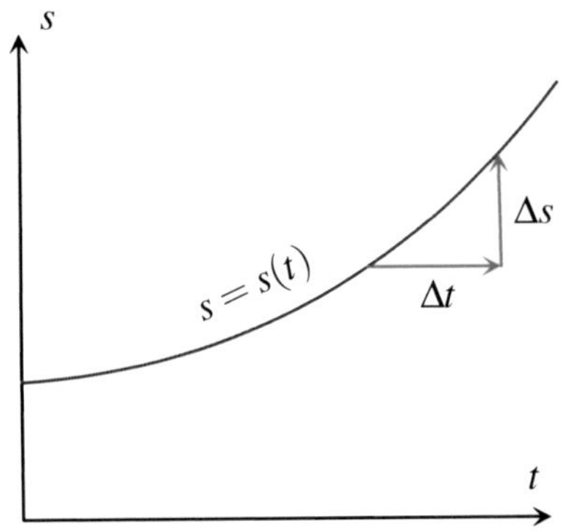

Figure 6.4: The characteristic in the (s,t)-plane

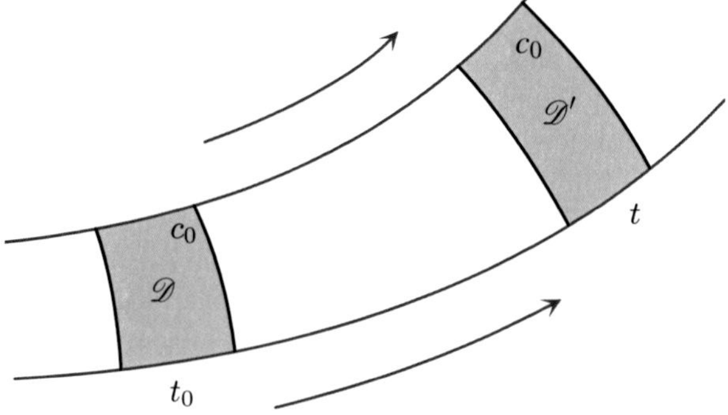

Figure 6.5: Plug flow

Chapter 7

Partial Differential Equations of the Second Order

We limit the discussion of second-order partial differential equations to linear equations and two-dimensional problems. Although the character of the partial differential equations remains much the same in three dimensions, the problems are much more difficult to solve in three-dimensional space.

Linear partial differential equations are defined as partial differential equations where the coefficients may be functions of the independent variables but not of the dependent ones. Linear partial differential equations of the first order are always of the same type as opposed to linear partial differential equations of the second order, where we distinguish three different types in much the same way as we distinguish three different types of second-order curves: elliptic, parabolic, and hyperbolic.

7.1 Types of Partial Differential Equations

Any partial differential equation of the second order can be represented as a set of two first-order partial differential equations:

$$\overset{1}{A}_1 \frac{\partial u}{\partial x} + \overset{1}{A}_2 \frac{\partial u}{\partial y} + \overset{1}{B}_1 \frac{\partial v}{\partial x} + \overset{1}{B}_2 \frac{\partial v}{\partial y} = \overset{1}{E} \tag{7.1}$$

$$\overset{2}{A}_1 \frac{\partial u}{\partial x} + \overset{2}{A}_2 \frac{\partial u}{\partial y} + \overset{2}{B}_1 \frac{\partial v}{\partial x} + \overset{2}{B}_2 \frac{\partial v}{\partial y} = \overset{2}{E}. \tag{7.2}$$

We write this in indicial notation, set $x_1 \equiv x$, $x_2 \equiv y$, and use the Einstein summation convention,

$$\overset{k}{A}_i \frac{\partial u}{\partial x_i} + \overset{k}{B}_i \frac{\partial v}{\partial x_i} = \overset{k}{E} \qquad (k = 1, 2;\ i = 1, 2). \tag{7.3}$$

Equation (7.3) represents (7.1) for $k = 1$, and (7.2) for $k = 2$. We investigate whether it is possible, as for the first order partial differential equation, to make combinations

O. D. L. Strack, *Applications of Vector Analysis and Complex Variables in Engineering*, https://doi.org/10.1007/978-3-030-41168-8_7

of the two equations to yield two new equations, each representing differentiation of u and v in the same direction. One such combination is:

$$\overset{k}{\lambda}\overset{k}{A}_i \frac{\partial u}{\partial x_i} + \overset{k}{\lambda}\overset{k}{B}_i \frac{\partial v}{\partial x_i} = \overset{k}{\lambda}\overset{k}{E}, \tag{7.4}$$

where summation is taken over k $(k = 1,2)$. We investigate if it is possible to choose $\overset{k}{\lambda}$ such that $\overset{k}{\lambda}\overset{k}{A}_i$ and $\overset{k}{\lambda}\overset{k}{B}_i$ are co-linear vectors, i.e. that

$$\overset{k}{\lambda}\overset{k}{B}_i = \mu \overset{k}{\lambda}\overset{k}{A}_i. \tag{7.5}$$

If this is possible, (7.4) becomes

$$\overset{k}{\lambda}\overset{k}{A}_i \left[\frac{\partial u}{\partial x_i} + \mu \frac{\partial v}{\partial x_i} \right] = \overset{k}{\lambda}\overset{k}{E}. \tag{7.6}$$

The dot product $\overset{k}{\lambda}\overset{k}{A}_i \frac{\partial u}{\partial x_i}$ represents the projection of $\frac{\partial u}{\partial x_i}$ on the vector $\overset{k}{\lambda}\overset{k}{A}_i$, which is the resultant of $\overset{1}{\lambda}\overset{1}{A}_i$ and $\overset{2}{\lambda}\overset{2}{A}_i$, multiplied by the magnitude of $\overset{k}{\lambda}\overset{k}{A}_i$. If we introduce the arc length s along the curve tangent to $\overset{k}{\lambda}\overset{k}{A}_i$, then the projection of $\frac{\partial u}{\partial x_i}$ onto the s direction equals the directional derivative $\frac{du}{ds}$. Hence, we can write (7.6) as:

$$\beta \left(\frac{du}{ds} + \mu \frac{dv}{ds} \right) = \overset{k}{\lambda}\overset{k}{E}, \tag{7.7}$$

where β is the length of the vector $\overset{k}{\lambda}\overset{k}{A}_i$:

$$\begin{aligned} \beta^2 &= (\overset{1}{\lambda}\overset{1}{A}_j + \overset{2}{\lambda}\overset{2}{A}_j)(\overset{1}{\lambda}\overset{1}{A}_j + \overset{2}{\lambda}\overset{2}{A}_j) \\ &= \overset{1}{\lambda}{}^2\overset{1}{A}_j\overset{1}{A}_j + 2\overset{1}{\lambda}\overset{2}{\lambda}\overset{1}{A}_j\overset{2}{A}_j + \overset{2}{\lambda}{}^2\overset{2}{A}_j\overset{2}{A}_j. \end{aligned} \tag{7.8}$$

Whether or not it is possible to write the system of equations (7.1) and (7.2) in the form (7.7) depends upon whether there are real numbers $\overset{k}{\lambda}$ and μ that satisfy (7.5). Since we may multiply (7.4) by any factor without affecting the solution, we may choose either $\overset{1}{\lambda}$ or $\overset{2}{\lambda}$ as 1. We choose

$$\overset{1}{\lambda} = \lambda \tag{7.9}$$

$$\overset{2}{\lambda} = 1, \tag{7.10}$$

and expand (7.5):

$$\lambda\overset{1}{B}_i + \overset{2}{B}_i = \mu(\lambda\overset{1}{A}_i + \overset{2}{A}_i) \qquad i = 1,2. \tag{7.11}$$

This represents a system of two equations with two unknowns. We solve for λ after eliminating μ by contracting both sides of the equation with $\varepsilon_{ij}(\lambda \overset{1}{A}_j + \overset{2}{A}_j)$, so that the right-hand side of the equation vanishes:

$$\varepsilon_{ij}\lambda^2 \overset{1}{B}_i \overset{1}{A}_j + \varepsilon_{ij}(\overset{1}{B}_i \overset{2}{A}_j + \overset{2}{B}_i \overset{1}{A}_j)\lambda + \varepsilon_{ij}\overset{2}{B}_i \overset{2}{A}_j = 0. \tag{7.12}$$

We multiply both sides by -1 and rename indices,

$$\varepsilon_{ij}\lambda^2 \overset{1}{A}_i \overset{1}{B}_j + \varepsilon_{ij}(\overset{1}{A}_i \overset{2}{B}_j + \overset{2}{A}_i \overset{1}{B}_j)\lambda + \varepsilon_{ij}\overset{2}{A}_i \overset{2}{B}_j = 0, \tag{7.13}$$

or

$$a\lambda^2 + b\lambda + c = 0, \tag{7.14}$$

where

$$\begin{aligned} a &= \overset{1}{A}_1 \overset{1}{B}_2 - \overset{1}{A}_2 \overset{1}{B}_1 \\ b &= \overset{1}{A}_1 \overset{2}{B}_2 - \overset{1}{A}_2 \overset{2}{B}_1 + \overset{2}{A}_1 \overset{1}{B}_2 - \overset{2}{A}_2 \overset{1}{B}_1 \\ c &= \overset{2}{A}_1 \overset{2}{B}_2 - \overset{2}{A}_2 \overset{2}{B}_1 . \end{aligned} \tag{7.15}$$

Equation (7.14) has real roots only if

$$b^2 - 4ac \geq 0 \tag{7.16}$$

There are three possibilities:

$$\begin{array}{lll} \text{case 1 (hyperbolic)} & b^2 - 4ac > 0: & \text{two characteristics} \\ \text{case 2 (parabolic)} & b^2 - 4ac = 0: & \text{two coinciding characteristics} \\ \text{case 3 (elliptic)} & b^2 - 4ac < 0: & \text{two complex characteristics} \end{array} \tag{7.17}$$

We consider these three types of second-order partial differential equations separately and in reverse order; first the elliptic case, second the parabolic case, and third the hyperbolic case. In the latter case, we have two characteristic directions, and two different roots, $\underset{1}{\lambda}$ and $\underset{2}{\lambda}$, of (7.13).

If there are two different roots for λ, then there are two roots for μ, and two equations of the form (7.5):

$$\underset{1}{\lambda}\overset{1}{B}_1 + \overset{2}{B}_1 = \underset{1}{\mu}(\underset{1}{\lambda}\overset{1}{A}_1 + \overset{2}{A}_1), \tag{7.18}$$

and

$$\underset{2}{\lambda}\overset{1}{B}_1 + \overset{2}{B}_1 = \underset{2}{\mu}(\underset{2}{\lambda}\overset{1}{A}_1 + \overset{2}{A}_1). \tag{7.19}$$

We obtain expressions for the two roots $\underset{1}{\lambda}$ and $\underset{2}{\lambda}$ from (7.14)

$$\underset{1}{\lambda}, \underset{2}{\lambda} = \frac{-b \pm \sqrt{b^2 - 4ac}}{2a}, \tag{7.20}$$

with a, b, and c given by (7.15), and we obtain the two corresponding values for $\underset{1}{\mu}$ and $\underset{2}{\mu}$ from (7.18) and (7.19):

$$\underset{1}{\mu} = \frac{\underset{1}{\lambda}\overset{1}{B}_1 + \overset{2}{B}_1}{\underset{1}{\lambda}\overset{1}{A}_1 + \overset{2}{A}_1}$$
$$\underset{2}{\mu} = \frac{\underset{2}{\lambda}\overset{1}{B}_1 + \overset{2}{B}_1}{\underset{2}{\lambda}\overset{1}{A}_1 + \overset{2}{A}_1}. \tag{7.21}$$

Equation (7.7) represents two equations, one for each of the two expressions for μ:

$$\frac{du}{ds_1} + \underset{1}{\mu}\frac{dv}{ds_1} = \underset{1}{c}$$
$$\frac{du}{ds_2} + \underset{2}{\mu}\frac{dv}{ds_2} = \underset{2}{c}, \tag{7.22}$$

where

$$\underset{j}{c} = \frac{\underset{j}{\lambda}\overset{1}{E} + \overset{2}{E}}{\underset{j}{\beta}} \qquad j = 1,2, \tag{7.23}$$

where β_j is given by (7.8),

$$\underset{j}{\beta} = \sqrt{\underset{j}{\lambda}^2 \overset{1}{A}_i\overset{1}{A}_i + 2\underset{j}{\lambda}\overset{1}{A}_i\overset{2}{A}_i + \overset{2}{A}_i\overset{2}{A}_i} \qquad j = 1,2. \tag{7.24}$$

The characteristic directions are parallel to the vectors $\underset{j}{\overset{k}{\lambda}}\overset{k}{A}_i = \underset{j}{\lambda}\overset{1}{A}_i + \overset{2}{A}_i$ $(j = 1,2)$; if $\underset{j}{\alpha}$ represents the angle between the j-characteristic and the x-axis, then

$$\tan \underset{j}{\alpha} = \frac{\underset{j}{\lambda}\overset{1}{A}_2 + \overset{2}{A}_2}{\underset{j}{\lambda}\overset{1}{A}_1 + \overset{2}{A}_1} \qquad j = 1,2. \tag{7.25}$$

We consider examples of partial differential equations in the next two sections to illustrate the process of determining the characteristic directions. The first example is the wave equation, which has two real characteristics. The second example is the diffusion equation, which has two coinciding characteristics.

7.2 Two Real Characteristics

The wave equation has two real characteristics. We study this equation in detail in Chapter 11, but here determine only its characteristic directions and the differential equations along the characteristics. The wave equation can be written in the form of two partial differential equations in terms of two dependent variables, u and v, whose meaning depends on the physical process that the equation represents,

$$\begin{aligned}\frac{\partial u}{\partial x} - \frac{\partial v}{\partial \tau} &= 0 \\ \frac{\partial u}{\partial \tau} - \frac{\partial v}{\partial x} &= 0,\end{aligned} \tag{7.26}$$

where x is a spatial coordinate and τ [L] is time multiplied by some factor of dimension [L/T].

The coefficients $\overset{k}{A}$ and $\overset{k}{B}$ are:

$$\begin{aligned}&\overset{1}{A}_1 = 1 \quad \overset{1}{A}_2 = 0 \quad \overset{1}{B}_1 = 0 \quad \overset{1}{B}_2 = -1 \quad \overset{1}{E} = 0 \\ &\overset{2}{A}_1 = 0 \quad \overset{2}{A}_2 = 1 \quad \overset{2}{B}_1 = -1 \quad \overset{2}{B}_2 = 0 \quad \overset{2}{E} = 0,\end{aligned} \tag{7.27}$$

We obtain expressions for a, b, and c from (7.15),

$$a = \overset{1}{A}_1 \overset{1}{B}_2 - \overset{1}{A}_2 \overset{1}{B}_1 = -1 \tag{7.28}$$

$$b = \overset{1}{A}_1 \overset{2}{B}_2 - \overset{1}{A}_2 \overset{2}{B}_1 + \overset{2}{A}_1 \overset{1}{B}_2 - \overset{2}{A}_2 \overset{1}{B}_1 = 0 \tag{7.29}$$

$$c = \overset{2}{A}_1 \overset{2}{B}_2 - \overset{2}{A}_2 \overset{2}{B}_1 = 1, \tag{7.30}$$

so that $b^2 - 4ac = 4$; the system is hyperbolic. The constants $\underset{1}{\lambda}$ and $\underset{2}{\lambda}$ are given by (7.20),

$$\underset{1}{\lambda} = -b + \frac{\sqrt{b^2 - 4ac}}{2a} = \frac{\sqrt{4}}{-2} = -1 \tag{7.31}$$

$$\underset{2}{\lambda} = -b - \frac{\sqrt{b^2 - 4ac}}{2a} = -\frac{\sqrt{4}}{-2} = 1, \tag{7.32}$$

and we find the constants $\underset{1}{\mu}$ and $\underset{2}{\mu}$ from (7.21), (7.31) and (7.32):

$$\underset{1}{\mu} = \frac{\underset{1}{\lambda}\overset{1}{B}_1 + \overset{2}{B}_1}{\underset{1}{\lambda}\overset{1}{A}_1 + \overset{2}{A}_1} = \frac{-1}{-1} = +1 \tag{7.33}$$

$$\underset{2}{\mu} = \frac{\underset{2}{\lambda}\overset{1}{B}_1 + \overset{2}{B}_1}{\underset{2}{\lambda}\overset{1}{A}_1 + \overset{2}{A}_1} = \frac{-1}{1} = -1. \tag{7.34}$$

We obtain expressions for the characteristic directions from (7.25),

$$\tan \underset{1}{\alpha} = \left[\frac{d\tau}{dx}\right]_1 = \frac{\underset{1}{\lambda}\overset{1}{A}_2 + \overset{2}{A}_2}{\underset{1}{\lambda}\overset{1}{A}_1 + \overset{2}{A}_1} = \frac{1}{-1}, \tag{7.35}$$

and

$$\tan \underset{2}{\alpha} = \frac{\underset{2}{\lambda} \overset{1}{A}_2 + \overset{2}{A}_2}{\underset{2}{\lambda} \overset{1}{A}_1 + \overset{2}{A}_1} = \frac{1}{+1}. \tag{7.36}$$

We derive the equation for the characteristic s_1 from (7.35),

$$\left(\frac{d\tau}{dx}\right)_1 = -1 \rightarrow \tau = -x + s_2 \qquad (s_1\text{-characteristic}); \tag{7.37}$$

s_2 is the constant of integration, constant only along the s_1-characteristic. The independent variables in the characteristic directions, s_1 and s_2, vary only in their coordinate directions. This is illustrated in Figure 7.1.

$$s_2 = -x - \tau \qquad s_1 - \text{characteristic}. \tag{7.38}$$

Similarly,

$$s_1 = x - \tau \qquad s_2 - \text{characteristic}. \tag{7.39}$$

We choose the signs of s_1 and s_2 according to the angles $\underset{1}{\alpha}$ and $\underset{2}{\alpha}$, which are chosen between 0 and π; this choice defines the positive directions of the new coordinates, see Figure 7.1. We transform the differential equations (7.26) along the characteristics by

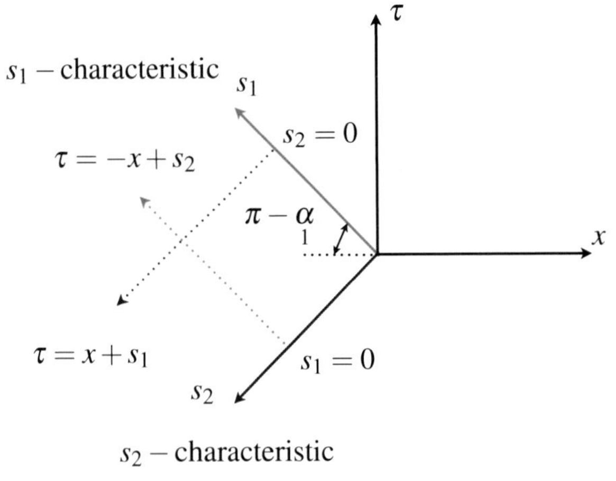

Figure 7.1: The coordinates along the characteristics.

the use of (7.22), with $\underset{j}{\mu} = \pm 1$,

$$\frac{du}{ds_1} + \frac{dv}{ds_1} = 0, \tag{7.40}$$

and

$$\frac{du}{ds_2} - \frac{dv}{ds_2} = 0. \tag{7.41}$$

These differential equations can be integrated:

$$u + v = \underset{1}{c} \qquad (\text{along } s_1), \tag{7.42}$$

and

$$u - v = \underset{2}{c} \qquad (\text{along } s_2). \tag{7.43}$$

The sum of u and v is constant along the s_1-characteristics, whereas their difference is constant along the s_2-characteristics. We obtain values for u and v at points of intersection of the s_1- and s_2-characteristics by solving (7.42) and (7.43) together.

The hyperbolic differential equation differs from the differential equations with either a single or complex characteristics in that discontinuities in the solution propagate from the boundary into the domain along characteristics.

We next analyze a parabolic differential equation, determine the characteristic, and interpret its meaning for the solutions.

7.3 Two Coinciding Characteristics

The diffusion equation is an example of a partial differential equation of the second order with two coinciding characteristics. We solve this equation for a number of applications in Chapter 10.

We determine the type of differential equation for the one-dimensional case of the diffusion equation:

$$\frac{\partial^2 c}{\partial x^2} = \frac{1}{\alpha} \frac{\partial c}{\partial t}. \tag{7.44}$$

We introduce a scaled time τ, $\tau = \alpha t$ with the dimension [L^2], so that

$$\frac{\partial^2 c}{\partial x^2} - \frac{\partial c}{\partial \tau} = 0, \tag{7.45}$$

and determine the system of two first-order partial differential equations equivalent to (7.45) by setting $\partial c / \partial x = u$, so that (7.45) becomes

$$\frac{\partial u}{\partial x} - \frac{\partial c}{\partial \tau} = 0. \tag{7.46}$$

The second equation is the definition of u,

$$\frac{\partial c}{\partial x} = u, \tag{7.47}$$

Equations(7.46) and (7.47) form a system of two partial differential equations of the first order, as considered in equations (7.1) and (7.2). The constants $\overset{k}{A}_i$ and $\overset{k}{B}_i$ are:

$$\begin{array}{lllll} \overset{1}{A}_1 = 1 & \overset{1}{A}_2 = 0 & \overset{1}{B}_1 = 0 & \overset{1}{B}_2 = -1 & \overset{1}{E} = 0 \\ \overset{2}{A}_1 = 0 & \overset{2}{A}_2 = 0 & \overset{2}{B}_1 = 1 & \overset{2}{B}_2 = \;\; 0 & \overset{2}{E} = u. \end{array} \tag{7.48}$$

We find expressions for a, b, and c from the latter equation and (7.15)

$$a = -1 \qquad b = 0 \qquad c^* = 0, \tag{7.49}$$

where we use c^* instead of c to avoid confusion with the concentration. It follows that $b^2 - 4ac^* = 0$: the system is parabolic. There is only one characteristic; the differential equation cannot be solved using the method of characteristics.

We determine the characteristic direction to investigate the behavior of solutions to this equation. The value for λ is

$$\lambda = 0, \tag{7.50}$$

indicating that the second equation is along the characteristic; (7.47) contains a deriva-

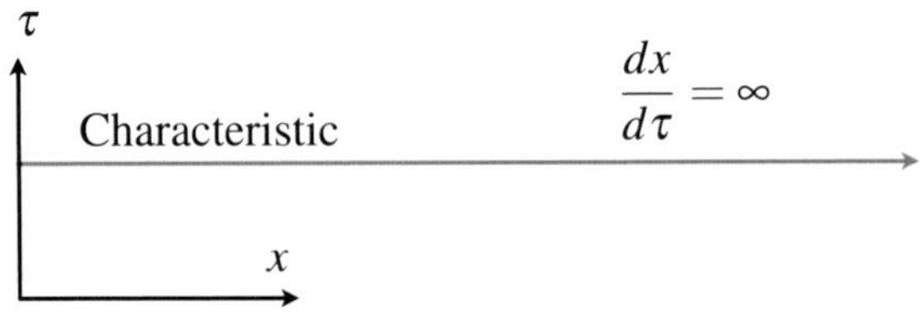

Figure 7.2: The case of a single characteristic; disturbances travel with infinite speed

tive with respect to x only. The lines τ =constant in the x, τ-space are the characteristics. Since discontinuities in c travel along the characteristics, they travel with infinite speed, because a line $\tau =$ constant in the x, τ-diagram means that $dx/d\tau = \infty$. This is illustrated in Figure 7.2. The choice (7.10), $\overset{2}{\lambda} = 1$, rules out the case $\overset{2}{\lambda} = 0$, which would occur if the first equation were written along the characteristic, i.e., if (7.47) were chosen as the first equation, and (7.46) as the second one.

7.4 Complex Characteristic Directions

In most textbooks complex characteristics are not considered to be useful for solving problems. We analyze the case of complex characteristics, rather than abandoning it out of hand, and find it useful; it leads to an approach that can be applied with considerable advantage over using real variables. The advantage of using complex variables to solve Laplace's equation in two dimensions is well-known, and amounts to using only one of the characteristics; this is possible because the characteristic is complex. Using both characteristics leads directly to a method of treating complex variables consistent with Wirtinger calculus, Wirtinger [1927]. We apply this approach in Chapter 8.

Chapter 8

The Elliptic Case: Two Complex Characteristics

Elliptic partial differential equations of the second order are relatively common in engineering practice. Examples include irrotational and divergence-free fluid flow, groundwater flow, and linear elasticity.

A class of problems can be written in terms of the divergence and curl of a vector field. We consider that both the divergence and the curl of this vector field, v_i, are functions of position; the vector components themselves may occur in the curl and divergence, but not their derivatives for the partial differential equation to be linear. The expressions for the curl and the divergence are

$$\frac{\partial v_x}{\partial x} + \frac{\partial v_y}{\partial y} = -\gamma, \tag{8.1}$$

and

$$-\frac{\partial v_x}{\partial y} + \frac{\partial v_y}{\partial x} = \delta. \tag{8.2}$$

The constants $\overset{k}{A}_i$ and $\overset{k}{B}_i$ for the system (8.1) and (8.2) are:

$$\overset{1}{A}_1 = 1 \qquad \overset{1}{A}_2 = \ 0 \qquad \overset{1}{B}_1 = 0 \qquad \overset{1}{B}_2 = 1 \qquad \overset{1}{E} = -\gamma \tag{8.3}$$

$$\overset{2}{A}_1 = 0 \qquad \overset{2}{A}_2 = -1 \qquad \overset{2}{B}_1 = 1 \qquad \overset{2}{B}_2 = 0 \qquad \overset{2}{E} = \delta. \tag{8.4}$$

We obtain the values of a, b, and c from (7.15),

$$a = 1 \qquad b = 0 \qquad c = 1, \tag{8.5}$$

so that $b^2 - 4ac = -4$; the system is elliptic. Discontinuities in elliptic partial differential equations cannot exist inside the domain, only at the boundaries; a discontinuity

O. D. L. Strack, *Applications of Vector Analysis and Complex Variables in Engineering*, https://doi.org/10.1007/978-3-030-41168-8_8

at the boundary cannot propagate into the domain because there are no real characteristics[1]. The characteristic directions are given by (7.25) on page 80 and are

$$\tan \underset{j}{\alpha} = \frac{\underset{j}{\lambda}\overset{1}{A}_2 + \overset{2}{A}_2}{\underset{j}{\lambda}\overset{1}{A}_1 + \overset{2}{A}_1} = \frac{-1}{\underset{j}{\lambda}}, \tag{8.6}$$

where $\underset{j}{\lambda}$ are given by

$$\underset{1}{\lambda}, \underset{2}{\lambda} = \frac{-b \pm \sqrt{b^2 - 4ac}}{2a} = \pm \mathrm{i}, \tag{8.7}$$

where i is the purely imaginary number with modulus 1, defined by the property:

$$\mathrm{i}^2 = -1. \tag{8.8}$$

We combine (8.6) with (8.7),

$$\tan \alpha_j = \frac{-1}{\pm \mathrm{i}}, \tag{8.9}$$

and obtain expressions for $\underset{k}{\mu}$ from (7.21)

$$\underset{k}{\mu} = \frac{1}{\underset{k}{\lambda}} = \frac{1}{\pm \mathrm{i}} = \mp \mathrm{i}. \tag{8.10}$$

The characteristic directions define the equations for the characteristics by

$$\frac{dy}{dx} = \frac{-1}{\pm \mathrm{i}}. \tag{8.11}$$

We integrate to obtain the orientation of the axes in the new coordinates as we did for the hyperbolic case:

$$\begin{aligned} \mathrm{i}y &= -x + z \\ -\mathrm{i}y &= -x + \bar{z}, \end{aligned} \tag{8.12}$$

where z is constant along the first characteristic and $\bar{z}$ is constant along the second one.

$$\begin{aligned} x - \mathrm{i}y &= \bar{z} \\ x + \mathrm{i}y &= z. \end{aligned} \tag{8.13}$$

The characteristic directions are complex quantities for the elliptical case, and coordinates along these characteristics are commonly written as z and $\bar{z}$, where $\bar{z}$ is called the complex conjugate of z. The coordinate transformation from x and y to $\bar{z}$ and z is similar to the transformation for the hyperbolic case, except that the new coordinates are complex. The transformation is illustrated in Figure 8.1, which is similar to Figure 7.1

[1] As we saw when considering first-order partial differential equations, certain discontinuities can propagate along characteristics.

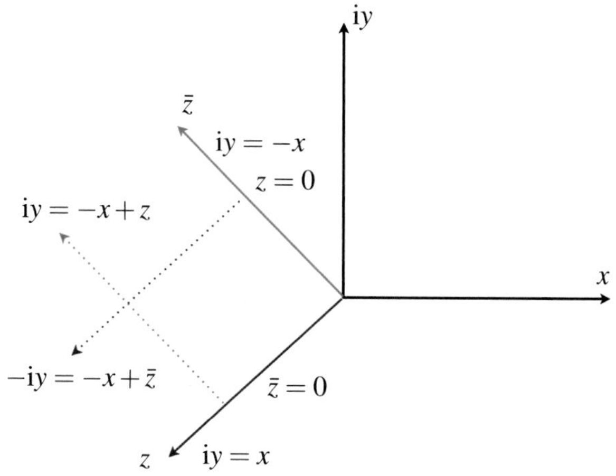

Figure 8.1: A diagram, illustrating the complex coordinates $\bar{z}$ and z

on page 82. This figure is intended for illustration only; it is an attempt to show a four dimensional relationship in a two-dimensional drawing, but may help because of its similarity to the case of two real characteristics.

We rewrite the partial differential equations (8.1) in terms of complex coordinates $\bar{z}$ and z. The $(\bar{z}, z)$-coordinate system is not Cartesian, and therefore vectors in terms of these coordinates behave differently than in Cartesian space. We do not divide the equations by the factors $\overset{k}{\beta}$, but instead use the expressions for $\underset{j}{\lambda}$ and $\overset{k}{A}_j$ in (7.6) on page 78:

$$\left[\underset{k}{\lambda}\overset{1}{A}_1 + \overset{2}{A}_1\right]\frac{\partial v_x}{\partial x} + \left[\underset{k}{\lambda}\overset{1}{A}_2 + \overset{2}{A}_2\right]\frac{\partial v_x}{\partial y} + \mu\left\{\left[\underset{k}{\lambda}\overset{1}{A}_1 + \overset{2}{A}_1\right]\frac{\partial v_y}{\partial x} + \left[\underset{k}{\lambda}\overset{1}{A}_2 + \overset{2}{A}_2\right]\frac{\partial v_y}{\partial y}\right\}$$
$$= \underset{k}{\lambda}\overset{1}{E} + \overset{2}{E}. \tag{8.14}$$

We use expressions (8.3), (8.4), (8.7), and (8.10) for $k = 1$,

$$\mathrm{i}\frac{\partial v_x}{\partial x} - \frac{\partial v_x}{\partial y} - \mathrm{i}\left[\mathrm{i}\frac{\partial v_y}{\partial x} - \frac{\partial v_y}{\partial y}\right] = -\mathrm{i}\gamma + \delta, \tag{8.15}$$

and divide both sides of the equation by i,

$$\frac{\partial v_x}{\partial x} + \mathrm{i}\frac{\partial v_x}{\partial y} - \mathrm{i}\left[\frac{\partial v_y}{\partial x} + \mathrm{i}\frac{\partial v_y}{\partial y}\right] = -(\gamma + \mathrm{i}\delta). \tag{8.16}$$

We observe from (8.13) that the full set of transformation equations is

$$\begin{aligned} \bar{z} &= x - \mathrm{i}y \qquad & x &= \frac{1}{2}(z + \bar{z}) \\ z &= x + \mathrm{i}y \qquad & y &= \frac{1}{2\mathrm{i}}(z - \bar{z}), \end{aligned} \tag{8.17}$$

and determine the partial derivatives of x and y with respect to $\bar{z}$:

$$\begin{aligned}\frac{\partial x}{\partial \bar{z}} &= \frac{1}{2}\\ \frac{\partial y}{\partial \bar{z}} &= -\frac{1}{2\mathrm{i}} = \frac{\mathrm{i}}{2}.\end{aligned} \tag{8.18}$$

The coordinates along the characteristics are not Cartesian, and we apply the formal transformation rules to determine the form of vectors, the gradient for example, in the non-Cartesian $(\bar{z}, z)$-space. We demonstrate in Section 8.3 that the rules for partial differentiation with respect to z and $\bar{z}$ apply, even though z and $\bar{z}$ are not independent variables; if $\bar{z}$ is known, so is z. Working with complex variables can be understood without first applying the formal transformation rules. It is up to the reader to decide whether to study the formal non-Cartesian coordinate transformation of the (x, y)-space to the $(\bar{z}, z)$-space in Section 8.3.

We write 1 as $2\partial x/\partial \bar{z}$ and i as $2\partial y/\partial \bar{z}$ in (8.16):

$$2\left[\frac{\partial v_x}{\partial x}\frac{\partial x}{\partial \bar{z}} + \frac{\partial v_x}{\partial y}\frac{\partial y}{\partial \bar{z}}\right] - 2\mathrm{i}\left[\frac{\partial v_y}{\partial x}\frac{\partial x}{\partial \bar{z}} + \frac{\partial v_y}{\partial y}\frac{\partial y}{\partial \bar{z}}\right] = -(\gamma + \mathrm{i}\delta). \tag{8.19}$$

The terms in the brackets represent $\partial v_x/\partial \bar{z}$ and $\partial v_y/\partial \bar{z}$, so that:

$$2\frac{\partial v_x}{\partial \bar{z}} - 2\mathrm{i}\frac{\partial v_y}{\partial \bar{z}} = -(\gamma + \mathrm{i}\delta). \tag{8.20}$$

We introduce new complex variables w and ω, defined as

$$\boxed{w = v_x - \mathrm{i}v_y}, \tag{8.21}$$

and

$$\boxed{\omega = \gamma + \mathrm{i}\delta}, \tag{8.22}$$

so that (8.20) simplifies to

$$\boxed{2\frac{\partial w}{\partial \bar{z}} = -\omega}. \tag{8.23}$$

We obtain the equation along the second characteristic, z, from (8.14) for $k = 2$ with $\underset{2}{\lambda} = -\mathrm{i}$ and $\underset{2}{\mu} = \mathrm{i}$ in a similar manner:

$$2\frac{\partial \bar{w}}{\partial z} = -\bar{\omega}. \tag{8.24}$$

This equation is the complex conjugate of (8.23), and does not provide additional information. It is the advantage of using complex variables that only one of the partial differential equations is required; only one of the two characteristics is needed to solve problems.

8.1 Boundary-Value Problems

Because the characteristics are imaginary, discontinuities cannot propagate into the domain; they can exist only at points of the boundary. Boundary values must be specified either for a single independent variable, or by a single condition from which a dependent variable can be determined. Boundary conditions must be stated for the entire boundary.

We understand this by considering the one-dimensional case where $v_y = 0$ and the governing equation reduces to (8.1). Since v_x can be represented as the derivative of some function F, (8.1) can be integrated, and two constants can be determined from the boundary values of either v_x, or F, one at some point $x = x_1$, and the other at a second point $x = x_2$. These two points together constitute the boundary for the one-dimensional case. For the general two-dimensional case, the boundary could be some closed contour around the domain, with one boundary condition specified along the entire boundary.

8.2 The Helmholtz Decomposition Theorem

The Helmholtz decomposition theorem states that the general vector field in two dimensions can be represented in terms of the sum of the gradient of a potential and the curl of a scalar stream function. This implies that, with the indices now ranging from 1 to 2,

$$v_i = -\partial_i \Phi - \varepsilon_{3ij} \partial_j \Psi, \tag{8.25}$$

where

$$-\partial_i v_i = \partial_i \partial_i \Phi = \nabla^2 \Phi = \gamma, \tag{8.26}$$

and

$$\varepsilon_{3ij} \partial_i v_j = -\varepsilon_{3ij} \partial_i \varepsilon_{3jm} \partial_m \Psi = \varepsilon_{3ij} \varepsilon_{3mj} \partial_i \partial_m \Psi. \tag{8.27}$$

We express the product of the epsilon tensors in terms of Kronicker deltas:

$$\varepsilon_{3ij} \varepsilon_{3mj} = \delta_{33} \delta_{im} - \delta_{3m} \delta_{3i} = \delta_{im}. \tag{8.28}$$

where δ_{3m} and δ_{3i} are zero as 3 is outside the range of i and j. We apply this to (8.27), writing ε_{3ij} as ε_{ij}:

$$\varepsilon_{ij} \partial_i v_j = \delta_{im} \partial_i \partial_m \Psi = \partial_i \partial_i \Psi = \nabla^2 \Psi, \tag{8.29}$$

so that:

$$\nabla^2 \Psi = \delta. \tag{8.30}$$

We express the complex derivative of Φ with respect to z in terms of the components of its gradient:

$$\frac{\partial \Phi}{\partial z} = \frac{\partial \Phi}{\partial x} \frac{\partial x}{\partial z} + \frac{\partial \Phi}{\partial y} \frac{\partial y}{\partial z} = \frac{1}{2} \left[\frac{\partial \Phi}{\partial x} - \mathrm{i} \frac{\partial \Phi}{\partial y} \right], \tag{8.31}$$

where we applied (8.17). We also write the complex derivative of Ψ in terms of the components of its gradient:

$$\frac{\mathrm{i}}{2} \left[\frac{\partial \Psi}{\partial x} - \mathrm{i} \frac{\partial \Psi}{\partial y} \right] = \mathrm{i} \frac{\partial \Psi}{\partial z} = \frac{1}{2} \left[\frac{\partial \Psi}{\partial y} + \mathrm{i} \frac{\partial \Psi}{\partial x} \right], \tag{8.32}$$

and write the general expression for v_i, (8.25), in complex form as

$$v_x - \mathrm{i}v_y = -\frac{\partial\Phi}{\partial x} - \varepsilon_{12}\frac{\partial\Psi}{\partial y} - \mathrm{i}\left[-\frac{\partial\Phi}{\partial y} - \varepsilon_{21}\frac{\partial\Psi}{\partial x}\right] = -\left[\frac{\partial\Phi}{\partial x} - \mathrm{i}\frac{\partial\Phi}{\partial y}\right] - \left[\frac{\partial\Psi}{\partial y} + \mathrm{i}\frac{\partial\Psi}{\partial x}\right], \tag{8.33}$$

use the complex velocity function w,

$$w = v_x - \mathrm{i}v_y, \tag{8.34}$$

and apply (8.31) and (8.32):

$$w = v_x - \mathrm{i}v_y = -\left[\frac{\partial\Phi}{\partial x} - \mathrm{i}\frac{\partial\Phi}{\partial y}\right] - \mathrm{i}\left[\frac{\partial\Psi}{\partial x} - \mathrm{i}\frac{\partial\Psi}{\partial y}\right] = -2\frac{\partial\Phi}{\partial z} - 2\mathrm{i}\frac{\partial\Psi}{\partial z}. \tag{8.35}$$

We combine the latter expression for w with (8.23),

$$-2\frac{\partial w}{\partial \bar{z}} = \gamma + \mathrm{i}\delta = 4\frac{\partial^2\Phi}{\partial z\partial\bar{z}} + 4\mathrm{i}\frac{\partial^2\Psi}{\partial z\partial\bar{z}}. \tag{8.36}$$

The operator $\partial^2/(\partial z\partial\bar{z})$ is real, so that we can separate real and imaginary parts:

$$4\frac{\partial^2\Phi}{\partial z\partial\bar{z}} = -\gamma \qquad 4\frac{\partial^2\Psi}{\partial z\partial\bar{z}} = \delta. \tag{8.37}$$

It follows from (8.26), (8.30), and (8.37) that the complex form of the Laplacian is:

$$\nabla^2 = 4\frac{\partial^2}{\partial z\partial\bar{z}}. \tag{8.38}$$

Recall that the derivations presented here are based on the assumption that partial differentiation applies to the complex variables z and $\bar{z}$. That this is indeed correct is shown in the next section, where the rules for non-Cartesian coordinate transformations are formally applied to complex variables, without using the common interpretation of partial differentiation, i.e., that the one independent variable is kept constant, while the other is varied.

The derivation given above is sufficient to follow most of the applications; Section 8.3 can be skipped.

8.3 Vectors and Tensors in Complex Space

We formally transform v_i by applying the formulas developed for non-Cartesian coordinate transformations in Section 5.3. Although the coordinate transformation contains the number i, the transformation still fulfills the criteria of uniqueness and reversibility.

8.3.1 The Bases in Complex Space

The complex coordinates are non-Cartesian; we must determine both the contravariant and covariant base vectors $\underset{m}{\tau}^k$ and $\overset{m}{\gamma}_k$ in order to apply the transformations. These base vectors are given by (5.31) and (5.32) on page 55,

$$\underset{m}{\tau}^{k} = \frac{\partial x_k}{\partial \tilde{x}_m} \tag{8.39}$$

$$\overset{m}{\gamma}_k = \frac{\partial \tilde{x}_m}{\partial x_k}, \tag{8.40}$$

where

$$\tilde{x}_m = (\bar{z}, z) \qquad x_m = (x, y). \tag{8.41}$$

The covariant base vectors are

$$\overset{1}{\gamma}_k = \frac{\partial \tilde{x}_1}{\partial x_i} = \left(\frac{\partial \bar{z}}{\partial x}, \frac{\partial \bar{z}}{\partial y}\right) = (1, -\mathrm{i}) \tag{8.42}$$

$$\overset{2}{\gamma}_k = \frac{\partial \tilde{x}_2}{\partial x_i} = \left(\frac{\partial z}{\partial x}, \frac{\partial z}{\partial y}\right) = (1, \mathrm{i}). \tag{8.43}$$

The second base vector $\overset{2}{\gamma}_k$ is related to the first base vector $\overset{1}{\gamma}_k$; it is the complex conjugate of $\overset{1}{\gamma}_k$,

$$\overset{2}{\gamma}_k = \overline{\overset{1}{\gamma}_k}. \tag{8.44}$$

We obtain similar expressions for the contravariant base vectors

$$\underset{1}{\tau}^{m} = \frac{\partial x_m}{\partial \tilde{x}_1} = \left(\frac{\partial x}{\partial \bar{z}}, \frac{\partial y}{\partial \bar{z}}\right) = \tfrac{1}{2}(1, \mathrm{i}) \tag{8.45}$$

$$\underset{2}{\tau}^{k} = \frac{\partial x_m}{\partial \tilde{x}_2} = \left(\frac{\partial x}{\partial z}, \frac{\partial y}{\partial z}\right) = \tfrac{1}{2}(1, -\mathrm{i}). \tag{8.46}$$

As for the covariant bases, the second contravariant base vector is the complex conjugate of the first one:

$$\underset{2}{\tau}^{k} = \overline{\underset{1}{\tau}^{k}}. \tag{8.47}$$

The covariant and contravariant base vectors are related:

$$\overset{m}{\gamma}_k = 2\overline{\underset{m}{\tau}^{k}}. \tag{8.48}$$

8.3.2 Transformation of Vectors

We use the expressions for the base vectors to transform vectors and tensors. We obtain the contravariant components from

$$\tilde{v}^1 = \overset{1}{\gamma}_j v_j = v_1 - \mathrm{i} v_2 \tag{8.49}$$

$$\tilde{v}^2 = \overset{2}{\gamma}_j v_j = v_1 + \mathrm{i} v_2, \tag{8.50}$$

and observe that, as for the independent variables $\bar{z}$ and z, the one component is the complex conjugate of the other. We use a single symbol, w, to represent both

$$w = \tilde{v}^1 = v_x - \mathrm{i}v_y \tag{8.51}$$

$$\bar{w} = \tilde{v}^2 = v_x + \mathrm{i}v_y, \tag{8.52}$$

and find the expressions for the covariant components of a vector in a similar fashion,

$$\tilde{v}_1 = \underset{1}{\tau}^j v_j = \tfrac{1}{2}(v_x + \mathrm{i}v_y) = \tfrac{1}{2}\overline{w} \tag{8.53}$$

$$\tilde{v}_2 = \underset{2}{\tau}^j v_j = \tfrac{1}{2}(v_x - \mathrm{i}v_y) = \tfrac{1}{2}w. \tag{8.54}$$

The modules of the vector, $v_i v_i$, is an invariant, which transforms as

$$v_i v_i = \tilde{v}_k \tilde{v}^k = \tfrac{1}{2}w\overline{w} + \tfrac{1}{2}w\overline{w} = w\overline{w}, \tag{8.55}$$

which indeed equals $v_x^2 + v_y^2$ as follows from (8.51). The differential operator is a covariant vector;

$$\tilde{\partial}_m = \left(\frac{\partial}{\partial \bar{z}}, \frac{\partial}{\partial z}\right), \tag{8.56}$$

and its contravariant components are:

$$\tilde{\partial}^m = 2\left(\frac{\partial}{\partial z}, \frac{\partial}{\partial \bar{z}}\right). \tag{8.57}$$

As for non-Cartesian coordinate systems, the natural way to write the gradient of a scalar function is in terms of its covariant components,

$$\frac{\partial F}{\partial \bar{z}} = \tfrac{1}{2}\left(\frac{\partial F}{\partial x} + \mathrm{i}\frac{\partial F}{\partial y}\right) \tag{8.58}$$

$$\frac{\partial F}{\partial z} = \tfrac{1}{2}\left(\frac{\partial F}{\partial x} - \mathrm{i}\frac{\partial F}{\partial y}\right). \tag{8.59}$$

We are now in a position to transform the divergence of the velocity vector v_i by contracting the covariant components of the gradient with the contravariant components of the velocity vector; this is necessary to obtain the invariant, i.e., to write an expression in terms of the new coordinates in such a way that it is equivalent to the corresponding one in the Cartesian system,

$$\partial_j v_j = \tilde{\partial}_j \tilde{v}^j = \frac{\partial w}{\partial \bar{z}} + \frac{\partial \bar{w}}{\partial z} = -\gamma. \tag{8.60}$$

8.3.3 Transformation of Tensors

We transform a tensor A_{mn} into a purely contravariant tensor:

$$\tilde{A}^{pq} = \overset{p}{\gamma}_m \overset{q}{\gamma}_n A_{mn}. \tag{8.61}$$

There are only two independent components of a tensor of the second rank; the component $\tilde{A}^{21}$ is the complex conjugate of the component $\tilde{A}^{12}$; we apply (8.44):

$$\tilde{A}^{21} = \overset{2}{\gamma}_m \overset{1}{\gamma}_n A_{mn} = \overline{\overset{1}{\gamma}_m \overset{2}{\gamma}_n A_{mn}} = \overline{\tilde{A}^{12}}. \tag{8.62}$$

The component $\tilde{A}^{22}$ is the complex conjugate of $\tilde{A}^{11}$:

$$\tilde{A}^{22} = \overset{2}{\gamma}_m \overset{2}{\gamma}_n A_{mn} = \overline{\overset{1}{\gamma}_m \overset{1}{\gamma}_n A_{mn}} = \overline{\tilde{A}^{11}}. \tag{8.63}$$

The purely covariant tensor components are related to the contravariant ones; we use (8.48),

$$\tilde{A}_{mn} = \underset{m}{\tau^p}\,\underset{n}{\tau^q} A_{pq} = \tfrac{1}{4}\overline{\overset{m}{\gamma}_p \overset{n}{\gamma}_q A_{pq}} = \tfrac{1}{4}\overline{\tilde{A}^{mn}}. \tag{8.64}$$

We obtain similar relations between the mixed transformed tensor components

$$\tilde{A}^{m}{}_{n} = \overset{m}{\gamma}_p \underset{n}{\tau^q} A_{pq} = \overline{\underset{m}{\tau^p}\overset{n}{\gamma}_q A_{pq}} = \overline{\tilde{A}_{m}{}^{n}}. \tag{8.65}$$

Only two independent tensor components exist, $\tilde{A}^{11}$ and $\tilde{A}^{12}$; we obtain all the other ones from these two by applying equations (8.61) through (8.65).

Problem 8.1
Demonstrate that $\tilde{A}^{1}{}_{1} = \tilde{A}_{21}$ and $\tilde{A}^{1}{}_{2} = \tilde{A}_{22}$.

8.3.4 A Combined Kronecker Delta and Alternating Tensor: The Combination Matrix

Working in complex space with vectorial properties such as divergence and curl, requires that the Kronecker delta and the epsilon tensor be transformed. There are four possible choices for each tensor; we begin by obtaining the transformed tensor $\tilde{\delta}^{m}{}_{n}$. Application of the transformation rule gives

$$\tilde{\delta}^{m}{}_{n} = \overset{m}{\gamma}_j \underset{n}{\tau^k}\delta_{jk} = \overset{m}{\gamma}_j \underset{n}{\tau^j} = \delta_{mn}. \tag{8.66}$$

We find, similarly

$$\tilde{\delta}_{m}{}^{n} = \underset{m}{\tau^j}\overset{n}{\gamma}_k \delta_{jk} = \underset{m}{\tau^j}\overset{n}{\gamma}_j = \delta_{mn}. \tag{8.67}$$

The transformed Kronecker delta with both components contravariant becomes

$$\tilde{\delta}^{mn} = \overset{m}{\gamma}_k \overset{n}{\gamma}_j \delta_{ij} = \overset{m}{\gamma}_k \overset{n}{\gamma}_k = g^{mn} = \begin{pmatrix} 0 & 2 \\ 2 & 0 \end{pmatrix}, \tag{8.68}$$

and with both components covariant:

$$\tilde{\delta}_{mn} = \underset{m}{\tau^k}\,\underset{n}{\tau^j}\delta_{ij} = \underset{m}{\tau^j}\,\underset{n}{\tau^j} = g_{mn} = \begin{pmatrix} 0 & \frac{1}{2} \\ \frac{1}{2} & 0 \end{pmatrix}. \tag{8.69}$$

The transformed Kronecker delta is called the metric tensor; the metric tensor equals the Kronecker delta for the case of mixed covariant/contravariant components.

We transform the epsilon tensor in a similar manner. Again, there are four possible combinations; we consider them in the same sequence as for the Kronecker delta. We obtain for $\tilde{\varepsilon}^m{}_n$

$$\tilde{\varepsilon}^m{}_n = \overset{m}{\gamma}_j \underset{n}{\tau}^k \varepsilon_{jk}. \tag{8.70}$$

The individual components are:

$$\begin{aligned} \tilde{\varepsilon}^1{}_1 &= \overset{1}{\gamma}_1 \underset{1}{\tau}^2 - \overset{1}{\gamma}_2 \underset{1}{\tau}^1 = \mathrm{i}/2 - (-\mathrm{i})/2 = \mathrm{i} \\ \tilde{\varepsilon}^1{}_2 &= \overset{1}{\gamma}_1 \underset{2}{\tau}^2 - \overset{1}{\gamma}_2 \underset{2}{\tau}^1 = -\frac{1}{2}\mathrm{i} + \frac{1}{2}\mathrm{i} = 0 \\ \tilde{\varepsilon}^2{}_2 &= \overline{\tilde{\varepsilon}^1{}_1} = -\mathrm{i}. \end{aligned} \tag{8.71}$$

or

$$\tilde{\varepsilon}^m{}_n = \begin{pmatrix} \mathrm{i} & 0 \\ 0 & -\mathrm{i} \end{pmatrix}. \tag{8.72}$$

We obtain the components of $\tilde{\varepsilon}_m{}^n$ similarly:

$$\tilde{\varepsilon}_m{}^n = \underset{m}{\tau}^j \overset{n}{\gamma}_k \varepsilon_{jk} = \overline{\tilde{\varepsilon}^m{}_n}, \tag{8.73}$$

which gives

$$\begin{aligned} \tilde{\varepsilon}_1{}^1 &= -\mathrm{i}/2 + (-\mathrm{i})/2 = -\mathrm{i} \\ \tilde{\varepsilon}_1{}^2 &= \tilde{\varepsilon}_2{}^1 = 0 \\ \tilde{\varepsilon}_2{}^2 &= \mathrm{i}, \end{aligned} \tag{8.74}$$

or

$$\tilde{\varepsilon}^n{}_m = \begin{pmatrix} -\mathrm{i} & 0 \\ 0 & \mathrm{i} \end{pmatrix}. \tag{8.75}$$

The purely covariant components of the epsilon tensor are

$$\tilde{\varepsilon}_{mn} = \underset{m}{\tau}^i \underset{n}{\tau}^j \varepsilon_{ij}, \tag{8.76}$$

which gives

$$\begin{aligned} \tilde{\varepsilon}_{11} &= \overline{\tilde{\varepsilon}_{22}} = \underset{1}{\tau}^1 \underset{1}{\tau}^2 - \underset{1}{\tau}^2 \underset{2}{\tau}^1 = 0 \\ \tilde{\varepsilon}_{12} &= \underset{1}{\tau}^1 \underset{2}{\tau}^2 - \underset{1}{\tau}^2 \underset{2}{\tau}^1 = -\frac{1}{4}\mathrm{i} - \frac{1}{4}\mathrm{i} = -\mathrm{i}/2 \\ \tilde{\varepsilon}_{21} &= \overline{\tilde{\varepsilon}_{12}} = \mathrm{i}/2, \end{aligned} \tag{8.77}$$

or

$$\tilde{\varepsilon}_{mn} = \begin{pmatrix} 0 & -\mathrm{i}/2 \\ \mathrm{i}/2 & 0 \end{pmatrix}. \tag{8.78}$$

The purely contravariant components $\tilde{\varepsilon}^{mn}$ are:

$$\tilde{\varepsilon}^{mn} = \overset{m}{\gamma}_k \overset{n}{\gamma}_j \varepsilon_{ij} = 4\overline{\tilde{\varepsilon}_{mn}}, \tag{8.79}$$

which gives

$$\begin{aligned} \tilde{\varepsilon}^{11} &= \tilde{\varepsilon}^{22} = 0 \\ \tilde{\varepsilon}^{12} &= 2\mathrm{i} \\ \tilde{\varepsilon}^{21} &= -2\mathrm{i}, \end{aligned} \tag{8.80}$$

or

$$\tilde{\varepsilon}^{mn} = \begin{pmatrix} 0 & 2\mathrm{i} \\ -2\mathrm{i} & 0 \end{pmatrix}. \tag{8.81}$$

The Combination Matrix

We construct a useful matrix by combining the Kronecker delta and the epsilon tensor, which we call the combination matrix. There are two possible representations for this matrix in Cartesian space, κ_{mn} and K_{mn}, that we define as:

$$\kappa_{mn} = \delta_{mn} - \mathrm{i}\varepsilon_{mn}, \tag{8.82}$$

and

$$K_{mn} = \delta_{mn} + \mathrm{i}\varepsilon_{mn} = \bar{\kappa}_{mn}. \tag{8.83}$$

We add to the transformed Kronecker delta the corresponding epsilon tensor times i for κ and $-\mathrm{i}$ for K. The mixed contravariant/covariant components are

$$\tilde{\kappa}^{m}{}_{n} = \begin{pmatrix} 2 & 0 \\ 0 & 0 \end{pmatrix} \quad \tilde{K}^{m}{}_{n} = \begin{pmatrix} 0 & 0 \\ 0 & 2 \end{pmatrix}, \tag{8.84}$$

the mixed covariant/contravariant ones are

$$\tilde{\kappa}_{m}{}^{n} = \begin{pmatrix} 0 & 0 \\ 0 & 2 \end{pmatrix} \quad \tilde{K}_{m}{}^{n} = \begin{pmatrix} 2 & 0 \\ 0 & 0 \end{pmatrix}, \tag{8.85}$$

and the purely contravariant components are

$$\tilde{\kappa}^{mn} = \begin{pmatrix} 0 & 4 \\ 0 & 0 \end{pmatrix} \quad \tilde{K}^{mn} = \begin{pmatrix} 0 & 0 \\ 4 & 0 \end{pmatrix}. \tag{8.86}$$

The components, finally, of the purely covariant components are

$$\tilde{\kappa}_{mn} = \begin{pmatrix} 0 & 0 \\ 1 & 0 \end{pmatrix} \quad \tilde{K}_{mn} = \begin{pmatrix} 0 & 1 \\ 0 & 0 \end{pmatrix}. \tag{8.87}$$

The combination matrices are not tensors; they do not obey the transformation rules (8.62) through (8.65), even though δ_{ij} and ε_{ij} do.

The transformed matrices have the unique property that only a single component is unequal to zero. This implies that the product of two complex numbers can be represented in the form of two vector components in complex space, in such a way that the indices of these components match the single non-zero component of the combination matrix. This means that the contraction of the two vectors with the combination matrix has a single contribution only, and thus represents two invariants, one equal to the real part, and the other imaginary part of the result. As an example, consider two complex numbers,

$$\begin{aligned} a &= \tilde{a}^1 = a_1 - \mathrm{i}a_2 \\ b &= \tilde{b}^2 = b_1 + \mathrm{i}b_2. \end{aligned} \tag{8.88}$$

These choices for a and b are such that they match $\tilde{K}_{mn}$, which has the sole non-zero component $\tilde{K}_{12}$, so that:

$$ab = \tilde{a}^1\tilde{b}^2 = \tilde{a}^m\tilde{b}^n\tilde{K}_{mn} = a_m b_n K_{mn} = a_m b_n \delta_{mn} + \mathrm{i}a_m b_n \varepsilon_{mn} = a_m b_m + \mathrm{i}\varepsilon_{mn} a_m b_n. \tag{8.89}$$

The product of these two complex numbers represents the complex combination of the dot product and the cross product of the vectors a_m and b_n.

We can transform the divergence and curl in a similar manner by applying the process in reverse. We write the complex combination of divergence and curl as:

$$\omega = -\partial_j v_j + \mathrm{i}\varepsilon_{mn}\partial_m v_n = -\kappa_{mn}\partial_m v_n. \tag{8.90}$$

The scalar ω is an invariant, and we transform it into the complex plane, taking care that summation takes place over indices in opposite positions

$$\omega = -\tilde{\kappa}^m{}_n\tilde{\partial}_m\tilde{v}^n = -2\frac{\partial w}{\partial \bar{z}}, \tag{8.91}$$

or

$$\boxed{2\frac{\partial w}{\partial \bar{z}} = -\omega = -(\gamma + \mathrm{i}\delta),} \tag{8.92}$$

where γ is minus the divergence, and δ the curl.

Thus, even though the invariant, being a combination of curl and divergence, is the sum of products of four different components, the transformed invariant that represents both divergence and curl consists of a single product of two components, w and $\partial/\partial\bar{z}$.

Problem 8.2

Derive expressions for all possible representations of the stress tensor σ_{ij} in complex space.

Problem 8.3

Write the equilibrium conditions for the stress tensor in complex space.

Problem 8.4

The derivations of equations (8.89) and (8.92) given in the text are not unique. Present as many alternative derivations as you can.

8.3.5 The Potential Φ

Recall that the gradient of a potential function Φ is irrotational. We transform $v_k = -\partial_k \Phi$,

$$w = v_x - \mathrm{i}v_y = -\left(\frac{\partial \Phi}{\partial x} - \mathrm{i}\frac{\partial \Phi}{\partial y}\right) = -2\frac{\partial \Phi}{\partial z}. \tag{8.93}$$

as follows from (8.59). We differentiate this with respect to $\bar{z}$ and multiply by 2:

$$2\frac{\partial w}{\partial \bar{z}} = -(\gamma + \mathrm{i}\delta) = -4\frac{\partial^2 \Phi}{\partial z \partial \bar{z}}. \tag{8.94}$$

The potential Φ is real; the right-hand side of (8.94) is also real, since $\partial^2/(\partial z \partial \bar{z})$ is a real operator because of the occurrence of both z and $\bar{z}$ in the derivative. Thus, we find:

$$\gamma = 4\frac{\partial^2 \Phi}{\partial z \partial \bar{z}} \qquad \delta = 0. \tag{8.95}$$

This result confirms our previous finding that the gradient of a scalar potential is an irrotational vector field. Furthermore, the Laplacian of the potential equals minus the divergence, so that

$$\nabla^2 \Phi = 4\frac{\partial^2 \Phi}{\partial z \partial \bar{z}}. \tag{8.96}$$

This is an important finding; we can integrate this expression for given γ, first with respect to z and then with respect to $\bar{z}$, or vice versa.

8.3.6 The Stream Function Ψ

We investigate in a similar manner the case that the velocity field is the curl of a scalar function, a stream function Ψ, i.e.,

$$v_j = -\varepsilon_{jk}\partial_k \Psi. \tag{8.97}$$

We rewrite this in complex form by applying the transformation

$$\tilde{v}^j = -\tilde{\varepsilon}^{jk}\tilde{\partial}_k \Psi. \tag{8.98}$$

This gives, for $j = 1$ and $k = 2$,

$$\tilde{v}^1 = -\tilde{\varepsilon}^{12}\tilde{\partial}_2 \Psi. \tag{8.99}$$

We set $\tilde{v}^1 = w$ and $\tilde{\partial}_2 = \partial/\partial z$, and use (8.81):

$$w = -2\mathrm{i}\frac{\partial \Psi}{\partial z}. \tag{8.100}$$

We obtain for $\omega = \gamma + \mathrm{i}\delta = -2\partial w/\partial \bar{z}$,

$$-2\frac{\partial w}{\partial \bar{z}} = 4\mathrm{i}\frac{\partial^2 \Psi}{\partial z \partial \bar{z}} = \gamma + \mathrm{i}\delta. \tag{8.101}$$

Thus, if the velocity field can be written in terms of the curl of a scalar stream function, in two dimensions, then the flow field is divergence-free and we obtain

$$\gamma = 0 \qquad \delta = 4\frac{\partial^2 \Psi}{\partial z \partial \bar{z}} = \nabla^2 \Psi. \tag{8.102}$$

8.3.7 The General Vector Field

We may write, in general

$$w = -2\frac{\partial \Gamma}{\partial z} \tag{8.103}$$

where Γ is a complex function of both z and $\bar{z}$

$$\Gamma(z,\bar{z}) = \Phi(z,\bar{z}) + \mathrm{i}\Psi(z,\bar{z}) \tag{8.104}$$

Problem 8.5

The derivation of equation (8.100) given in the text is not unique. Present as many alternative derivations as you can.

8.3.8 Helmholtz's Decomposition Theorem

A vector field that can be expressed entirely in terms of the gradient of a potential is irrotational, and a vector field that can be expressed entirely in terms of the curl of a stream function is divergence-free. What remains to be shown is that Helmholtz's decomposition theorem is true, i.e., that any vector field can be written as the sum of an irrotational and a divergence-free one. We show this by considering two vector fields. The actual vector field $\underset{a}{w}$ and a vector field $\underset{h}{w}$ defined as

$$\underset{h}{w} = -2\frac{\partial \Phi}{\partial z} - 2\mathrm{i}\frac{\partial \Psi}{\partial z}, \tag{8.105}$$

and chosen such that it satisfies:

$$\left[\frac{\partial^2 \Phi}{\partial x^2} + \frac{\partial^2 \Phi}{\partial y^2}\right] + \mathrm{i}\left[\frac{\partial^2 \Psi}{\partial x^2} + \frac{\partial^2 \Psi}{\partial y^2}\right] = 4\left[\frac{\partial^2 \Phi}{\partial z \partial \bar{z}} + \mathrm{i}\frac{\partial^2 \Psi}{\partial z \partial \bar{z}}\right] = \omega = (\gamma + \mathrm{i}\delta), \tag{8.106}$$

where ω equals $(\gamma + \mathrm{i}\delta)$ of the actual vector field so that

$$\frac{\partial \underset{a}{w}}{\partial \bar{z}} = \frac{\partial \underset{h}{w}}{\partial \bar{z}}. \tag{8.107}$$

We integrate this with respect to $\bar{z}$ and obtain the following relationship

$$\underset{a}{w} = \underset{h}{w} + h(z), \tag{8.108}$$

where $h(z)$ is a function of only z, called a holomorphic function. Thus, the representation (8.106) determines the actual vector field except for a holomorphic function. The latter function can be determined from the boundary conditions, i.e., Helmholtz's decomposition theorem holds: any vector field can be decomposed into an irrotational and a divergence-free one, provided that an appropriate vector field be added that is both irrotational and divergence-free, and chosen such that the combination of the three fields meets the boundary conditions.

8.3.9 Areal Integration

We examine the complex formulation of areal integrals, and find that significant simplification results as compared to a formulation in terms of Cartesian real coordinates. An infinitesimal area $dx_1 dx_2$ can be represented by the cross product of two vectors, one in each of the coordinate directions. These vectors are $\delta_{1p} dx_1 = (dx_1, 0)$ and $\delta_{2q} dx_2 = (0, dx_2)$, so that the infinitesimal area is expressed as

$$dA = \varepsilon_{mn} \delta_{1m} \delta_{2n} dx_1 dx_2. \tag{8.109}$$

We transform the area onto the $(\bar{z}, z)$ domain. In doing so, we must remember that the incremental vector dz^m is covariant; its components are in the coordinate directions. It is common practice to write these components as dz_m, but we will, in this instance, adhere to the precise notation dz^m. Both $\delta_{1m} dx_1$ and $\delta_{2m} dx_2$ represent vectors in the 1 and the 2 directions, respectively. The expression for dA is an invariant, and we write its equivalent form in the $\bar{z}, z$ space as

$$dA = \tilde{\varepsilon}_{mn} \tilde{\delta}_1{}^m \tilde{\delta}_2{}^n dz^1 dz^2. \tag{8.110}$$

Note the position of the summation indices, as required to obtain an invariant. The tensor $\tilde{\delta}_m{}^n$ is equal to the tensor δ_{mn}:

$$dA = \tilde{\varepsilon}_{12} dz d\bar{z}, \tag{8.111}$$

or, with (8.77)

$$dA = -\frac{\mathrm{i}}{2} dz d\bar{z}. \tag{8.112}$$

8.3.10 Integral Theorems

The integral theorems of Gauss and Stokes presented in Chapter 4 for three dimensions, concern the areal integration of the divergence and the curl, respectively. In two dimensions, these integral theorems can be derived by integrating the complex combination of divergence and curl, ω, over an area, using (8.111)

$$I = \iint\limits_{\mathscr{A}} (\gamma + \mathrm{i}\delta)\, dxdy = \iint\limits_{\mathscr{A}} \omega(z,\bar{z}) dxdy = \frac{1}{2\mathrm{i}} \iint\limits_{\mathscr{A}} \omega(z,\bar{z}) dz d\bar{z}. \tag{8.113}$$

We use (8.92) to replace $\omega = \gamma + \mathrm{i}\delta$ by $-2\partial w / \partial \bar{z}$,

$$I = \iint\limits_{\mathscr{A}} \omega dxdy = -\frac{1}{\mathrm{i}} \iint\limits_{\mathscr{A}} \frac{\partial w}{\partial \bar{z}} d\bar{z} dz = -\frac{1}{\mathrm{i}} \int\limits_{\mathscr{B}} w(z,\bar{z}) dz = \mathrm{i} \int\limits_{\mathscr{B}} w(z,\bar{z}) dz, \tag{8.114}$$

where $\mathscr{B}$ represents the boundary of $\mathscr{A}$. The integral represents minus the divergence plus i times the curl, integrated over the area A.

The term under the single integral sign in the trailing term in (8.114) is the product of two vector components; these components are:

$$w = \tilde{v}^1, \tag{8.115}$$

and

$$dz = d\tilde{x}^2. \tag{8.116}$$

Thus the product wdz is the product of two contravariant components. In order to identify the invariant this represents in Cartesian space (recall that the combination matrix shows that we can create the product of two complex vector components to represent some invariant in Cartesian space), we must find the transformed combination matrix that has a non-zero 1-2 component. The only one of the matrices shown in (8.84) through (8.87) with this property is $\tilde{K}_{12}$ in (8.87); we write wdz as

$$wdz = \tilde{K}_{12}\tilde{v}^1 d\tilde{x}^2 = \tilde{K}_{mn}\tilde{v}^m d\tilde{x}^n = K_{mn}v_m dx_n = v_m dx_m + \mathrm{i}\varepsilon_{mn}v_m dx_n. \tag{8.117}$$

The real part of wdz equals the dot product of v_m and the incremental tangential vector dx_m; this is the projection of v_m onto the tangential incremental vector, pointing to the left with respect to the outward normal. The term $\varepsilon_{mn}dx_n$ represents an incremental vector that points out of $\mathscr{A}$ and normal to dx_n, i.e., normal to $\mathscr{B}$ (see Figure 8.2), and is obtained from dx_n by rotating it clockwise over an angle $\pi/2$. The dot product $v_m\varepsilon_{mn}dx_n$ represents the dot product of v_m and the vector of length dx_n and normal to $\mathscr{B}$. The incremental vector $\varepsilon_{mn}dx_n$ points out of the domain and $\varepsilon_{mn}v_m dx_n$ is the

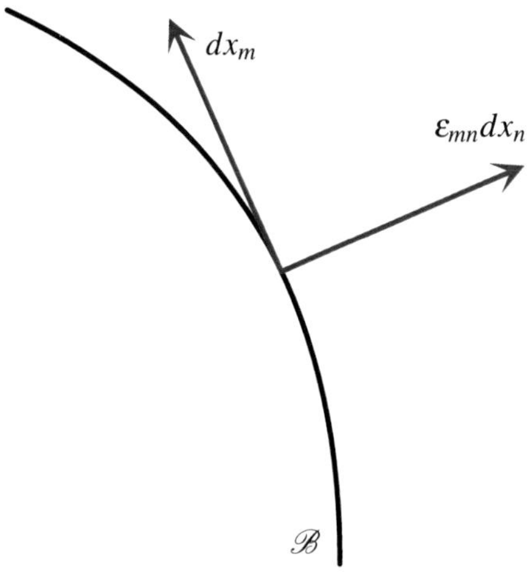

Figure 8.2: The incremental vectors dx_m and $\varepsilon_{mn}dx_m$

projection of v_k onto the outward normal multiplied by the increment ds along the boundary, i.e.,

$$\varepsilon_{mn}v_m dx_n = \underset{n}{v}ds, \tag{8.118}$$

where $\underset{n}{v}$ is the outward normal component of v_m. Similarly, $v_m dx_m$ is the projection of v_m onto the incremental vector along the boundary, taken positive pointing in the direction the contour is traveled, i.e., with the area $\mathscr{A}$ to the left. Hence

$$v_m dx_m = \underset{t}{v}ds, \tag{8.119}$$

where $\underset{t}{v}$ is the tangential component of v_m. We use (8.114) with (8.118) and (8.119) to obtain the boundary integral in terms of the normal and tangential components of v_k,

$$\iint_{\mathscr{A}} (\gamma + \mathrm{i}\delta) dxdy = \mathrm{i} \int_{\mathscr{B}} w dz = \mathrm{i} \int_{\mathscr{B}} \left(\underset{t}{v} + \mathrm{i} \underset{n}{v} \right) ds = \int_{\mathscr{B}} \left(-\underset{n}{v} + \mathrm{i} \underset{t}{v} \right) ds. \tag{8.120}$$

We separate this integral into real and imaginary parts and obtain Gauss's integral theorem, also known as the divergence theorem, for the real part, and Stokes's integral theorem for the imaginary part,

$$\iint_{\mathscr{A}} -\gamma dxdy = \iint_{\mathscr{A}} \partial_k v_k dxdy = \int_{\mathscr{B}} \underset{n}{v} ds \tag{8.121}$$

$$\iint_{\mathscr{A}} \delta dxdy = \iint_{\mathscr{A}} \varepsilon_{mn} \partial_m v_n dxdy = \int_{\mathscr{B}} \underset{t}{v} ds. \tag{8.122}$$

This relationship is illustrated in Figure 8.3.

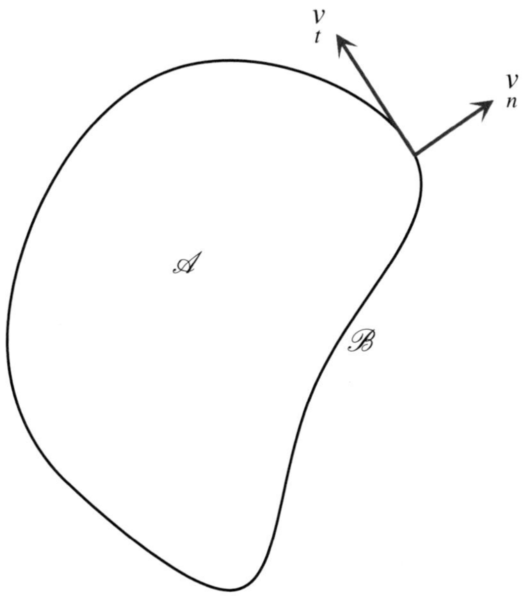

Figure 8.3: Integration of $\partial/\partial\bar{z}$ over an area $\mathscr{A}$ bounded by a contour $\mathscr{B}$.

Problem 8.6

Consider the functions

$$\omega = \gamma_0 + \mathrm{i}\delta_0, \tag{8.123}$$

and

$$\omega = az\bar{z}^2 + \mathrm{i}bz^2\bar{z}. \tag{8.124}$$

and the area $\mathscr{A}$ bounded by a circle $\mathscr{C}$, centered at the origin and of radius R.

Questions:

- Determine the function w that has the divergence and curl given by ω.
- Determine the areal integral of ω over area $\mathscr{A}$ by integration.
- Determine the same integral, but now by applying (8.114).
- Determine the tangential and normal components of w along $\mathscr{C}$.

8.4 Irrotational and Divergence-Free Vector Fields

If the vector field is both irrotational and divergence-free, (8.23) reduces to

$$\frac{\partial w}{\partial \bar{z}} = 0, \tag{8.125}$$

so that w is a function of z only, i.e.,

$$w = w(z). \tag{8.126}$$

Because the flow is both irrotational and divergence-free, we can represent the complex velocity function either as the gradient of a real potential, or as the curl of a real scalar stream function. If we express w in terms of a velocity potential, we write the gradient of the potential in terms of derivatives with respect to the complex variables z and $\bar{z}$

$$\begin{aligned} w = v_x - \mathrm{i}v_y = -\left[\frac{\partial \Phi}{\partial x} - \mathrm{i}\frac{\partial \Phi}{\partial y}\right] = & - \left[\frac{\partial \Phi}{\partial z}\frac{\partial z}{\partial x} + \frac{\partial \Phi}{\partial \bar{z}}\frac{\partial \bar{z}}{\partial x}\right] \\ & + \mathrm{i}\left[\frac{\partial \Phi}{\partial z}\frac{\partial z}{\partial y} + \frac{\partial \Phi}{\partial \bar{z}}\frac{\partial \bar{z}}{\partial y}\right], \end{aligned} \tag{8.127}$$

or, with $\partial z/\partial x = 1, \partial z/\partial y = \mathrm{i}$ and $\partial \bar{z}/\partial x = 1, \partial \bar{z}/\partial y = -\mathrm{i}$

$$w = -2\frac{\partial \Phi}{\partial z}. \tag{8.128}$$

As an alternative, we write the velocity vector as minus the curl of a scalar stream function,

$$v_k = -\varepsilon_{ij}\partial_j \Psi, \tag{8.129}$$

so that

$$w = v_x - \mathrm{i}v_y = -\frac{\partial \Psi}{\partial y} - \mathrm{i}\frac{\partial \Psi}{\partial x} = -\mathrm{i}\left[\frac{\partial \Psi}{\partial x} - \mathrm{i}\frac{\partial \Psi}{\partial y}\right]. \tag{8.130}$$

Comparing the latter equation with (8.127) and (8.128), we see that we may express w in terms of the derivative of Ψ with respect to z,

$$w = -2\mathrm{i}\frac{\partial \Psi}{\partial z}. \tag{8.131}$$

The representations of w in terms of the derivatives of Φ and Ψ with respect to z gives us several possibilities to represent w:

$$w = w(z) = -2\frac{\partial \Phi}{\partial z} = -2\mathrm{i}\frac{\partial \Psi}{\partial z} = -\left(\frac{\partial \Phi}{\partial z} + \mathrm{i}\frac{\partial \Psi}{\partial z}\right). \tag{8.132}$$

The final expression in the latter equation distributes the contribution to w equally between a real potential and a real stream function. We observe from the third and fourth expressions in the latter equation that

$$\frac{\partial \Phi}{\partial z} = \mathrm{i}\frac{\partial \Psi}{\partial z}, \tag{8.133}$$

and take the complex conjugate of both sides of this equation. The real functions Φ and Ψ are not affected, but i changes to $-\mathrm{i}$ and z to $\bar{z}$:

$$\frac{\partial \Phi}{\partial \bar{z}} = -\mathrm{i}\frac{\partial \Psi}{\partial \bar{z}}. \tag{8.134}$$

Equations (8.133) and (8.134) are equivalent complex forms of the Cauchy-Riemann equations; the latter equations are obtained in real form by separating the real and imaginary parts of these equations; both equations lead to the same two real equations:

$$\begin{aligned} \frac{\partial \Phi}{\partial x} &= \frac{\partial \Psi}{\partial y} \\ \frac{\partial \Phi}{\partial y} &= -\frac{\partial \Psi}{\partial x}. \end{aligned} \tag{8.135}$$

Integration of (8.134) with respect to $\bar{z}$ gives

$$\Phi + \mathrm{i}\Psi = f(z), \tag{8.136}$$

where $f(z)$ is a function of z only, which we call the complex potential and represent by the letter Ω,

$$\Omega = \Omega(z) = \Phi + \mathrm{i}\Psi. \tag{8.137}$$

We represent w as minus the derivative with respect to z of this complex potential:

$$w = -\frac{d\Omega}{dz}. \tag{8.138}$$

This is an ordinary derivative; Ω is a function of z only, in contrast to both the potential Φ and the stream function Ψ, which are real functions and must be functions of both z and $\bar{z}$.

We observe from (8.135) that the gradients of the velocity potential and the stream function are mutually orthogonal; their dot product is zero. Since the gradient of the velocity potential is parallel to the velocity, the gradient of the stream function is at right angles to the flow. Both the contours $\Psi = const$ and the velocity are normal to the gradient of the stream function; the velocity vector must be tangent to the contours $\Psi = const$.

8.4.1 Summary

The real and imaginary parts of a complex potential $\Omega(z)$, which is a function of z only, i.e., does not contain $\bar{z}$ in its expression, produces a complex derivative $d\Omega/dz$, which is minus the complex form of a velocity field, w. This velocity field is both divergence-free and irrotational. The real and imaginary parts of the complex potential are harmonic functions, i.e., they satisfy Laplace's equation, as follows from (8.95) and (8.102).

8.5 Functions of a Single Complex Variable

Functions of a single complex variable that can be differentiated an infinite number of times in a certain region are called *analytic* or *holomorphic* inside that region. We use the term holomorphic to identify this property for functions of a single complex variable; we use the term analytic if the function depends on both z and $\bar{z}$ and can be differentiated an infinite number of times with respect to either of the independent variables z and $\bar{z}$. Note that we call these variables independent, which they are only in the strict sense of a coordinate transformation; the special property of the number i, however, makes it possible to determine the second variable once the first is known.

We allow the divergence to be non-zero at isolated points and along lines, and possibly inside isolated domains in the area considered. In such cases the stream function cannot be single-valued, and will be singular at isolated points. For very special cases rotation may be permitted at isolated points, along lines, and in isolated domains. In this case the potential function will not be single-valued at the points or singular at isolated points. A singularity of a holomorphic function is defined as a point where the function cannot be differentiated an infinite number of times without producing an infinite value at the singular point.

Working with complex variables requires certain formulas to understand the meaning of the various forms of complex functions. We present a brief summary of such formulas below without giving formal proof of the equations; the reader is referred to the multitude of texts on this topic.

The complex variable z can be represented in terms of Cartesian coordinates (x,y) and in terms of radial coordinates (r,θ),

$$z = x + \mathrm{i}y = r(\cos\theta + \mathrm{i}\sin\theta). \tag{8.139}$$

A useful formula was developed by Euler and is obtained from the Taylor series expansion of the exponential function of $(\mathrm{i}\theta)$ about $\theta = 0$:

$$\mathrm{e}^{\mathrm{i}\theta} = 1 + \mathrm{i}\theta - \frac{\theta^2}{2!} - \mathrm{i}\frac{\theta^3}{3!} + \frac{\theta^4}{4!} + \mathrm{i}\frac{\theta^5}{5!} + \cdots. \tag{8.140}$$

We separate the real and imaginary parts:

$$\Re\mathrm{e}^{\mathrm{i}\theta} = 1 - \frac{\theta^2}{2!} + \frac{\theta^4}{4!} + \cdots, \tag{8.141}$$

and

$$\Im e^{i\theta} = \theta - \frac{\theta^3}{3!} + \frac{\theta^5}{5!} + \cdots. \tag{8.142}$$

The series expansions of the real and imaginary parts of $e^{i\theta}$ are the cosine and the sine,

$$e^{i\theta} = \cos\theta + i\sin\theta. \tag{8.143}$$

This formula, Euler's formula, permits us to write the polar form of z in a compact way:

$$z = |z|e^{i\theta}, \tag{8.144}$$

where $|z|$ is the modulus of z with

$$|z|^2 = z\bar{z} = x^2 + y^2. \tag{8.145}$$

Special values of $e^{i\theta}$

Euler's formula yields some special values for $e^{i\theta}$:

$$\begin{aligned} e^{i\pi/2} &= \cos\frac{\pi}{2} + i\sin\frac{\pi}{2} = i \\ e^{i\pi} &= \cos\pi + i\sin\pi = -1 \\ e^{-i\pi/2} &= \cos\frac{\pi}{2} + i\sin -\frac{\pi}{2} = -i. \end{aligned} \tag{8.146}$$

8.6 Basic Complex Functions

We present in this section a summary of complex functions including representation in the form of infinite series.

8.6.1 The Taylor Series

Any function that is holomorphic inside a circle of radius R can be expanded in terms of a Taylor series about the center of the circle, z_0,

$$f(z) = \sum_{n=0}^{\infty} a_n (z - z_0)^n, \tag{8.147}$$

where the coefficients a_n can be expressed in terms of the derivatives of $f(z)$ computed at z_0:

$$a_n = \frac{f^n(z_0)}{n!}. \tag{8.148}$$

We verify this by differentiating $f(z)$ multiple times and then setting $z = z_0$. This series converges for $(z - z_0)\overline{(z - z_0)} \leq R^2$. Points where not all derivatives of $f(z)$ exist are singular points, and may exist only outside the circle of convergence of $f(z)$. The Taylor series expansion of $f(z)$ about z_0 converges to the function $f(z)$ inside the circle centered at z_0. The theorem works both ways: if a function can be expanded in terms of a series (8.147), then the function is holomorphic inside the circle of convergence. All derivatives of $f(z)$ with respect to z exist inside the circle of convergence, as follows from (8.148).

8.6.2 The Asymptotic Expansion

The second example of a complex function is the asymptotic expansion, which contains only negative powers of z,

$$f(z) = \sum_{j=0}^{\infty} a_{-j}(z - z_0)^{-j}. \tag{8.149}$$

This expansion converges outside a circle with a radius such that it does not contain any singularities of $f(z)$ outside the circle. The Taylor series and the asymptotic expansion are the complex functions that are holomorphic in a domain bounded by a circle; one represents the holomorphic function inside the circle, and the other one the general holomorphic function outside a circle. The coefficients in the series expansions can be determined from the boundary conditions.

8.6.3 The Laurent Series

The holomorphic function in a domain bounded by two circles is the sum of the Taylor series and an asymptotic expansion; this series is called the Laurent series:

$$f(z) = \sum_{n=-\infty}^{\infty} a_n(z - z_0)^n. \tag{8.150}$$

8.7 Cauchy Integrals

A holomorphic function can be represented as a function of a single complex variable, z. Since its derivatives with respect to $\bar{z}$ are zero, all the derivatives of a holomorphic function are themselves holomorphic functions. We can write a function Ω that is holomorphic inside area $\mathscr{A}$ as the derivative of another function that is holomorphic inside $\mathscr{A}$, say F , i.e.,

$$\Omega = \frac{dF}{dz}. \tag{8.151}$$

We write the integral of Ω along the boundary $\mathscr{C}$ of $\mathscr{A}$:

$$\oint_{\mathscr{C}} \Omega dz = F(z_e) - F(z_s), \tag{8.152}$$

where z_s and z_e are the complex coordinates of the starting and end points of the integral, which are the same. The integral is zero, provided that $F(z)$ is single-valued; the resulting formula is the Cauchy-Goursat integral theorem,

$$\oint_{\mathscr{C}} \Omega dz = 0. \tag{8.153}$$

Another formula applicable to holomorphic functions is the Cauchy integral theorem,

$$\Omega(z) = \frac{1}{2\pi i} \oint_{\mathscr{C}} \frac{\Omega(\delta)}{\delta - z} d\delta \qquad z \in \mathscr{A}, \tag{8.154}$$

where z represents a point inside $\mathscr{A}$. We demonstrate that this formula is valid by rewriting the integral,

$$\frac{1}{2\pi\mathrm{i}}\oint\limits_{\mathscr{C}}\frac{\Omega(\delta)}{\delta-z}d\delta=\frac{1}{2\pi\mathrm{i}}\oint\limits_{\mathscr{C}}\frac{\Omega(\delta)-\Omega(z)}{\delta-z}d\delta+\frac{1}{2\pi\mathrm{i}}\oint\limits_{\mathscr{C}}\frac{\Omega(z)}{\delta-z}d\delta. \tag{8.155}$$

Since Ω is holomorphic, it can be expanded in a Taylor series about $\delta = z$:

$$\frac{1}{2\pi\mathrm{i}}\oint\frac{\Omega(\delta)}{\delta-z}d\delta=\frac{1}{2\pi\mathrm{i}}\oint\frac{\Omega(z)-\Omega(z)+\Omega'(z)(\delta-z)+\cdots}{\delta-z}d\delta+\frac{1}{2\pi\mathrm{i}}\oint\frac{\Omega(z)}{\delta-z}d\delta. \tag{8.156}$$

The factor $(\delta-z)$ in the term $\Omega'(z)(\delta-z)$ cancels the term $1/(\delta-z)$ in the first integral to the right of the equal sign; this is the integral of a holomorphic function and is zero,

$$\frac{1}{2\pi\mathrm{i}}\oint\frac{\Omega(\delta)}{\delta-z}d\delta=\frac{1}{2\pi\mathrm{i}}\Omega(z)\oint\limits_{\mathscr{C}}\frac{1}{\delta-z}d(\delta-z)=\frac{1}{2\pi\mathrm{i}}\Omega(z)\ln(\delta-z)|_{\mathscr{C}}\qquad z\in\mathscr{A}. \tag{8.157}$$

We separate the logarithm into its real and imaginary parts by the use of (8.144):

$$\ln(\delta-z)=\ln\left(|\delta-z|\mathrm{e}^{\mathrm{i}\theta}\right)=\ln|\delta-z|+\mathrm{i}\theta. \tag{8.158}$$

The real part of the logarithm $\ln(\delta-z)$ is the same at the starting and end points of the contour, but the imaginary part, θ, increases by 2π upon completion of the path of integration, see Figure 8.4, provided that point z is in area $\mathscr{A}$ so that:

$$\frac{1}{2\pi i}\oint\frac{\Omega(\delta)}{\delta-z}d\delta=\frac{1}{2\pi i}\Omega(z)2\pi i=\Omega(z). \tag{8.159}$$

If point z is outside area $\mathscr{A}$ the integral is zero, i.e.,

$$\frac{1}{2\pi\mathrm{i}}\oint\limits_{\mathscr{C}}\frac{\Omega(\delta)}{\delta-z}d\delta=0\qquad z\in\mathscr{A}^-, \tag{8.160}$$

where $\mathscr{A}^-$ represents the domain outside of area $\mathscr{A}$.

We differentiate the Cauchy integral (8.154) repeatedly to obtain integral expressions for the derivatives of Ω:

$$\frac{d^m\Omega(z)}{dz^m}=m!\frac{1}{2\pi\mathrm{i}}\oint\limits_{\mathscr{C}}\frac{\Omega(\delta)}{(\delta-z)^{m+1}}d\delta\qquad z\in\mathscr{A}. \tag{8.161}$$

The Cauchy integral allows us to obtain the coefficients of the Taylor series expansion of the function $\Omega(z)$ about the origin, $z=0$, in terms of the values of Ω along $\mathscr{C}$; these coefficients involve the derivatives of Ω at $z=0$. We set $z=0$ in (8.161) and choose for $\mathscr{C}$ the unit circle

$$\frac{1}{m!}\frac{d^m\Omega(0)}{dz^m}=\frac{1}{2\pi\mathrm{i}}\oint\limits_{\mathscr{C}}\frac{\Omega(\delta)}{\delta^{m+1}}d\delta. \tag{8.162}$$

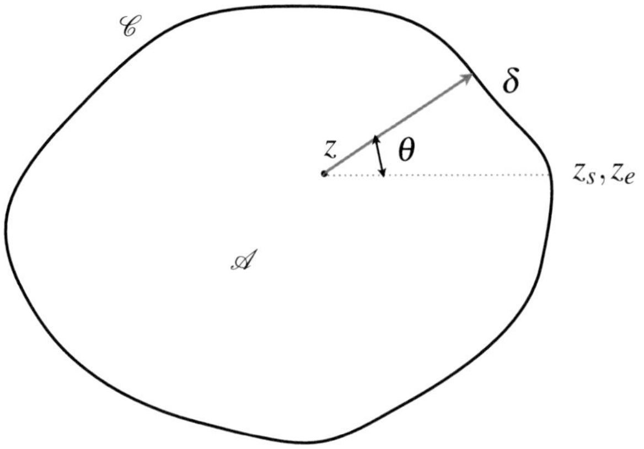

Figure 8.4: The argument of the term $(\delta - z)$ for z inside the domain

Since Ω is holomorphic in $\mathscr{A}$, it can be expanded in terms of a Taylor series with coefficients $a_m, m = 0, 1, \cdots$, and we obtain with (8.147) and (8.148)

$$a_m = \frac{1}{2\pi \mathrm{i}} \oint_{\mathscr{C}} \frac{\Omega(\delta)}{\delta^{m+1}} d\delta \qquad m = 0, 1, \cdots. \tag{8.163}$$

This equation expresses the coefficients of the Taylor expansion of Ω in terms of its values along the circle $\mathscr{C}$. In general, these complex boundary values are not known; the complex potential inside a domain is fully determined if either the real or imaginary part of Ω is given along $\mathscr{C}$ but not both. We address this problem by considering the integral

$$\frac{1}{2\pi \mathrm{i}} \oint_{\mathscr{C}} \Omega(\delta)\delta^n d\delta = 0 \qquad n \geq 0. \tag{8.164}$$

This integral is zero because the integrand is holomorphic in $\mathscr{A}$. The complex conjugate of a complex expression that is zero is also zero, i.e.,

$$\frac{1}{2\pi \mathrm{i}} \oint_{\mathscr{C}} \overline{\Omega(\delta)}\bar{\delta}^n d\bar{\delta} = 0 \qquad \delta \in \mathscr{C} \qquad n \geq 0. \tag{8.165}$$

We have that $\bar{\delta} = 1/\delta$ along the circle $\mathscr{C}$ (recall that $\mathscr{C}$ is the unit circle, i.e., $\delta\bar{\delta} = 1$), and we obtain

$$\frac{1}{2\pi \mathrm{i}} \oint_{\mathscr{C}} \overline{\Omega(\delta)} \frac{1}{\delta^{n+2}} d\delta = 0 \qquad \delta \in \mathscr{C} \qquad n \geq 0. \tag{8.166}$$

We choose $n = m - 1$ for $m > 1$ and add (8.166) to (8.163),

$$a_m = \frac{1}{2\pi \mathrm{i}} \oint_{\mathscr{C}} \frac{\Omega(\delta) + \overline{\Omega(\delta)}}{\delta^{m+1}} d\delta \qquad m = 1, \cdots \qquad \delta \in \mathscr{C}. \tag{8.167}$$

This equation yields the coefficients of the Taylor series expansion of Ω (z) about $z = 0$ in terms of the boundary values of the real part of the complex potential, Φ , i.e.

$$a_m = \frac{1}{\pi \mathrm{i}} \oint_{\mathscr{C}} \frac{\Phi(\delta)}{\delta^{m+1}} d\delta \qquad m = 1, \cdots . \tag{8.168}$$

Alternatively, we can subtract (8.166) from (8.163) which gives

$$a_m = \frac{1}{2\pi \mathrm{i}} \oint_{\mathscr{C}} \frac{\Omega(\delta) - \overline{\Omega(\delta)}}{\delta^{m+1}} d\delta \qquad m = 1, \cdots \qquad \delta \in \mathscr{C}. \tag{8.169}$$

Since $\Omega - \bar{\Omega} = 2\mathrm{i}\Psi$, this becomes:

$$a_m = \frac{1}{\pi} \oint_{\mathscr{C}} \frac{\Psi(\delta)}{\delta^{m+1}} d\delta \qquad m = 1, \cdots \qquad \delta \in \mathscr{C}. \tag{8.170}$$

This equation expresses the coefficients a_m in terms of the boundary values of the imaginary part of Ω. The special case that $m = 0$ is covered by setting m in (8.163) equal to zero, which gives

$$a_0 = \frac{1}{2\pi \mathrm{i}} \oint_{\mathscr{C}} \frac{\Omega(\delta)}{\delta} d\delta \qquad \delta \in \mathscr{C}. \tag{8.171}$$

Along the circle $\mathscr{C}$ we write δ as

$$\delta = \mathrm{e}^{\mathrm{i}\theta}, \tag{8.172}$$

and substitute this expression for δ in (8.171):

$$\begin{aligned} a_0 &= \frac{1}{2\pi \mathrm{i}} \oint_{\mathscr{C}} \Omega(\delta) \mathrm{e}^{-\mathrm{i}\theta} d\mathrm{e}^{\mathrm{i}\theta} = \frac{1}{2\pi \mathrm{i}} \oint_{\mathscr{C}} \Omega(\delta) \mathrm{e}^{-\mathrm{i}\theta} \mathrm{i} \mathrm{e}^{\mathrm{i}\theta} d\theta \\ &= \frac{1}{2\pi} \int_0^{2\pi} \Omega(\delta) d\theta \qquad \delta \in \mathscr{C}. \end{aligned} \tag{8.173}$$

We rewrite (8.168) in terms of θ :

$$a_m = \frac{1}{\pi \mathrm{i}} \int_0^{2\pi} \frac{\Phi(\delta)}{e^{\mathrm{i}(m+1)\theta}} d e^{\mathrm{i}\theta} = \frac{1}{\pi} \int_0^{2\pi} \Phi(\delta) \mathrm{e}^{-\mathrm{i}m\theta} d\theta \qquad m = 1, \cdots \qquad \delta \in \mathscr{C}, \tag{8.174}$$

and (8.170) becomes

$$a_m = \frac{\mathrm{i}}{\pi} \int_0^{2\pi} \Psi(\delta) \mathrm{e}^{-\mathrm{i}m\theta} d\theta \qquad m = 1, \cdots \qquad \delta \in \mathscr{C}. \tag{8.175}$$

These expressions for the coefficients of the series correspond to those obtained for a complex Fourier series.

Chapter 9

Applications of Complex Variables

We apply complex variables in this chapter to three topics taken from engineering practice, in the order of presentation: fluid mechanics, groundwater flow, and linear elasticity, as applied to soil mechanics and rock mechanics.

9.1 Flow of an in-viscid fluid

The material presented in this section is a continuation of the treatment of fluid mechanics in Chapter 3, Section 3.7, and serves to illustrate the application of complex variables in engineering. We begin by considering irrotational and divergence-free flow. The general solution to such problems in two dimensions is a function that is holomorphic, with the possible exception of isolated points. We present a few elementary functions, and later combine these to represent more general velocity fields.

9.1.1 Uniform Flow

A uniform velocity field has components $\underset{0}{v}_x$ and $\underset{0}{v}_y$ in the x- and y-directions that are constant in both time and space. The corresponding complex velocity function $w = v_x - \mathrm{i}v_y$ is:

$$w = \underset{0}{w} = \underset{0}{v}_x - \mathrm{i}\underset{0}{v}_y = \underset{0}{v}\mathrm{e}^{-\mathrm{i}\alpha} = -\frac{d\Omega}{dz}, \tag{9.1}$$

where v_0 is the magnitude of the velocity vector, and α its orientation. We integrate to obtain an expression for the complex potential:

$$\Omega = -\underset{0}{w}z = (-\underset{0}{v}_x + \mathrm{i}\underset{0}{v}_y)(x + \mathrm{i}y). \tag{9.2}$$

We have chosen the constant of integration to be zero; it does not affect the velocity field. We separate the complex potential into its real and imaginary parts:

$$\Omega = \Phi + \mathrm{i}\Psi = -\underset{0}{v}_x x - \underset{0}{v}_y + \mathrm{i}(-\underset{0}{v}_x y + \underset{0}{v}_y x), \tag{9.3}$$

O. D. L. Strack, *Applications of Vector Analysis and Complex Variables in Engineering*, https://doi.org/10.1007/978-3-030-41168-8_9

so that

$$\begin{aligned} \Phi &= -v_{x0}x - v_{y0}y \\ \Psi &= -v_{x0}y + v_{y0}x \end{aligned} \tag{9.4}$$

Recall that the stream function is constant along streamlines. We use the stream function for uniform flow to demonstrate that the difference in value of the stream function on different streamlines is equal in magnitude to the flow between these streamlines. Consider flow in the x-direction only, i.e., $v_{y0} = 0$. The stream function reduces to:

$$\Psi = -v_{x0}y, \tag{9.5}$$

so that indeed Ψ is constant along the streamlines, which are in the x-direction. The difference $\Delta\Psi$ in value of the stream function between a streamline at $y = y_1$ and a streamline $y = y_2$ is:

$$\Delta\Psi = \Psi(y_1) - \Psi(y_2) = -v_{x0}(y_1 - y_2) = -v_{x0}\Delta y. \tag{9.6}$$

The stream function increases in the negative y-direction. The velocity potential increases in the negative x-direction; the directions of increase of the velocity potential and the stream function correspond to a right-hand coordinate system, as shown in Figure 9.1.

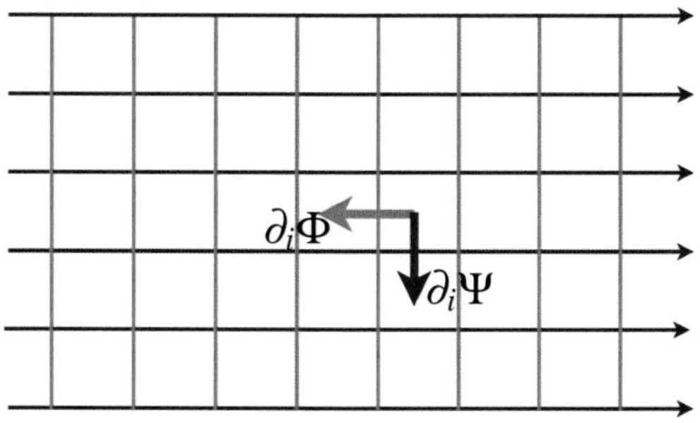

Figure 9.1: The stream function for uniform flow

9.1.2 The Point Sink

The second example is a point sink of discharge Q at $z = 0$,

$$\Omega = \Phi + \mathrm{i}\Psi = \frac{Q}{2\pi}\ln z. \tag{9.7}$$

We differentiate this with respect to z to obtain the complex velocity, w,

$$w = -\frac{d\Omega}{dz} = -\frac{Q}{2\pi z}, \tag{9.8}$$

and obtain the individual components by writing this in terms of radial coordinates r, θ:

$$w = v_x - \mathrm{i}v_y = |w|(\cos\alpha - \mathrm{i}\sin\alpha) = |w|\mathrm{e}^{-\mathrm{i}\alpha} = -\frac{Q}{2\pi}\frac{1}{r}\mathrm{e}^{-\mathrm{i}\theta}, \tag{9.9}$$

so that $\alpha = \theta$; the velocity vector points in the direction of $-r$; it is normal to any circle centered at the point sink.

We use (8.144) on page 105 to separate real and imaginary parts of the complex potential,

$$\Omega = \Phi + \mathrm{i}\Psi = \frac{Q}{2\pi}\ln[|z|e^{i\theta}] = \frac{Q}{2\pi}[\ln|z| + \mathrm{i}\theta], \tag{9.10}$$

The stream function is constant along lines $\theta = const$, i.e., along straight lines through the point sink. The stream function jumps by an amount Q across the negative real axis, assuming that the angle θ is taken between $-\pi$ and π; it is $-\pi$ for $y = 0^-$, $x \leq 0$, with 0^- indicating points just below the real axis, and π for $y = 0$, $x \leq 0$. The negative real axis is called the branch cut. The branch cut is a mathematical consequence of the presence of the point sink, which removes water from the system. Since the stream function records the amount of flow passing between streamlines, it is unavoidable that the stream function jumps somewhere; beginning at $x < 0$, $y = 0^-$, and continuing in counterclockwise direction around the point sink until we arrive at $x < 0$, $y = 0$, an amount Q will have passed between the streamlines along $y = 0^-$ and $y = 0$, hence the jump of Q of the stream function across the negative real axis. The branch cut mimics flow from the point sink to infinity (the jump in stream function across the cut corresponds to flow from the point sink to infinity); it exists to satisfy the condition of continuity of flow. This condition can be met by any shape of branch cut that does not intersect itself, and does connect the point sink to infinity. The complex potential for a point sink at a point z_w, rather than at the origin, is

$$\Omega = \frac{Q}{2\pi}\ln(z - z_w). \tag{9.11}$$

9.1.3 The Vortex

The complex potential for a vortex is the same function as that for a point sink, but the coefficient is purely imaginary, rather than real:

$$\Omega = \frac{A}{2\pi\mathrm{i}}\ln z \qquad \Im A = 0. \tag{9.12}$$

The complex velocity is

$$w = ve^{-\mathrm{i}\alpha} = -\frac{d\Omega}{dz} = -\frac{A}{2\pi\mathrm{i}z} = -\frac{A}{2\pi}e^{-\mathrm{i}\pi/2}\frac{1}{z} = -\frac{A}{2\pi}\frac{1}{|z|}\mathrm{e}^{-\mathrm{i}\theta-\mathrm{i}\pi/2}. \tag{9.13}$$

The velocity vector is tangent to circles centered at the vortex. The real and imaginary parts of the complex potential are:

$$\Omega = \Phi + \mathrm{i}\Psi = \frac{A}{2\pi}\theta - \mathrm{i}\frac{A}{2\pi}\ln|z|. \tag{9.14}$$

The stream function is constant along circles, whereas the potential is constant along radii emanating from the vortex. The potential is double-valued along the branch cut where θ jumps, usually taken along $\theta = \pm\pi$.

9.1.4 The Dipole

A dipole is the limiting case of a point sink of negative strength and one of positive strength approaching one another, with their discharges growing indefinitely, see Figure 9.2. We consider a point source at $z = \delta = d\mathrm{e}^{\mathrm{i}\beta}$ and a point sink at $z = -\delta = -d\mathrm{e}^{\mathrm{i}\beta}$, with a discharge $Q = s/(2d)$:

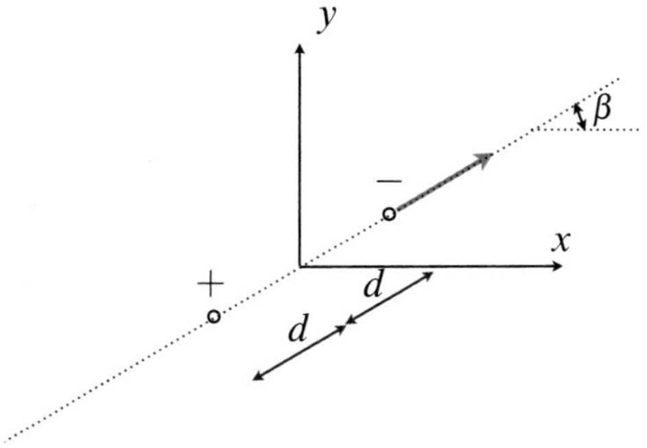

Figure 9.2: A dipole as the limiting case of a point sink and a point source

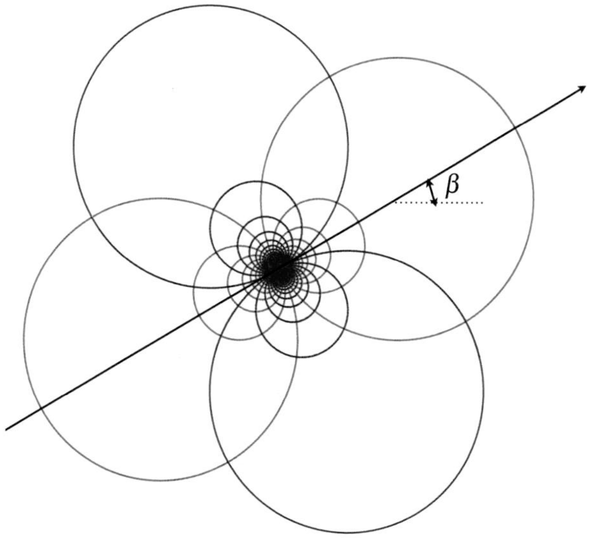

Figure 9.3: Equipotentials (in red) and streamlines (in blue) for a dipole oriented at an angle β to the x-axis

$$\Omega = -\lim_{\substack{d\to 0\\ Q\to\infty}} \frac{Q}{2\pi} \ln \frac{z-\delta}{z+\delta} = -\lim_{\substack{d\to 0\\ Q\to\infty}} \frac{Q}{2\pi} \ln \frac{1-\delta/z}{1+\delta/z} = -\lim_{\substack{d\to 0\\ Q\to\infty}} \frac{Q}{2\pi} \left[-\frac{2\delta}{z}\right]; \tag{9.15}$$

we expanded $\ln(1+x)$ about $x=0$. We use that $\delta = d\mathrm{e}^{\mathrm{i}\beta}$ and $Q = s/(2d)$,

$$\Omega = \lim_{\substack{d\to 0\\ Q\to\infty}} \frac{Q}{2\pi}\left[-\frac{2\delta}{z}\right] = \lim_{d\to 0} \frac{s}{4d\pi}\frac{2d\mathrm{e}^{\mathrm{i}\beta}}{z} = \frac{s}{2\pi}\frac{\mathrm{e}^{\mathrm{i}\beta}}{z}. \tag{9.16}$$

The complex potential for a dipole at an arbitrary point z_d rather than at the origin is:

$$\Omega = \frac{s}{2\pi}\frac{\mathrm{e}^{\mathrm{i}\beta}}{z - z_d}. \tag{9.17}$$

A plot of equipotentials and streamlines for a dipole oriented at an angle $\beta = \pi/6$ is shown in Figure 9.3.

9.2 Superposition of Elementary Solutions

The governing equations are linear; we can combine solutions to obtain new ones. The first example of superposition is that of flow around an impermeable cylinder.

9.2.1 Flow Around an Impermeable Cylinder

We determine both the velocity field and the pressure field for the case of flow around an impermeable cylinder of radius R in a field of uniform flow with velocity $\underset{0}{v}, \Im \underset{0}{v} = 0$, oriented at an angle α to the x-direction. The complex potential is

$$\Omega = -\left[\underset{0}{w}z + \underset{0}{\bar{w}}\frac{R^2}{z}\right], \tag{9.18}$$

where $\underset{0}{w}$ is the constant complex velocity

$$\underset{0}{w} = \underset{0}{v}(\cos\alpha - \mathrm{i}\sin\alpha) = \underset{0}{v}\mathrm{e}^{-\mathrm{i}\alpha}. \tag{9.19}$$

We verify that the boundary condition along the cylinder wall is met by setting $z\bar{z} = R^2$,

$$\Omega = \Phi + \mathrm{i}\Psi = -\left[\underset{0}{w}z + \underset{0}{\bar{w}}\frac{R^2\bar{z}}{\bar{z}z}\right] = -\left[\underset{0}{w}z + \underset{0}{\bar{w}}\frac{R^2}{R^2}\bar{z}\right] = -(\underset{0}{w}z + \underset{0}{\bar{w}}\bar{z}). \tag{9.20}$$

This expression is real; Ψ is zero along the boundary, i.e., it is a streamline. We determine an expression for the complex velocity by differentiating the complex potential:

$$w = -\frac{d\Omega}{dz} = \underset{0}{w} - \underset{0}{\bar{w}}\frac{R^2}{z^2}. \tag{9.21}$$

The velocity field approaches that for uniform flow, $w \to \underset{0}{w}$, for $z \to \infty$, as the term R^2/z approaches zero.

We consider the special case that $\alpha = 0$,

$$\underset{0}{w} = \underset{0}{v} \qquad \Im \underset{0}{v} = 0, \tag{9.22}$$

and compute the magnitude of the velocity along the cylinder wall by setting z equal to $R\mathrm{e}^{\mathrm{i}\theta}$ in (9.21) after replacing both $\underset{0}{w}$ and $\underset{0}{\bar{w}}$ by $\underset{0}{v}$ and using (9.21),

$$w = \underset{0}{v}\left[1 - \frac{R^2}{R^2}\mathrm{e}^{-2\mathrm{i}\theta}\right] = \underset{0}{v}(1 - \mathrm{e}^{-2\mathrm{i}\theta}) = \underset{0}{v}\mathrm{e}^{-\mathrm{i}\theta}(\mathrm{e}^{\mathrm{i}\theta} - \mathrm{e}^{-\mathrm{i}\theta}) = 2\underset{0}{v}\mathrm{i}\mathrm{e}^{-\mathrm{i}\theta}\sin\theta. \tag{9.23}$$

We write w as $|w|\mathrm{e}^{-\mathrm{i}\beta}$, where β is the angle between the velocity vector and the real axis,

$$|w|\mathrm{e}^{-\mathrm{i}\beta} = 2\underset{0}{v}\mathrm{e}^{\mathrm{i}\pi/2}\mathrm{e}^{-\mathrm{i}\theta}\sin\theta = 2\underset{0}{v}\mathrm{e}^{-\mathrm{i}(\theta-\pi/2)}\sin\theta, \tag{9.24}$$

so that the orientation of the velocity, given by β, is

$$\sin\theta = |\sin\theta| \qquad \beta = \theta - \pi/2 \qquad 0 \le \theta \le \pi \tag{9.25}$$

$$\sin\theta = |\sin\theta|\mathrm{e}^{\mathrm{i}\pi} \qquad \beta = \theta + \pi/2 \qquad -\pi \le \theta \le 0. \tag{9.26}$$

The velocity vector is tangential to the cylinder wall and points to the right with respect to the outward normal to the wall when $\sin\theta$ is positive; otherwise it points to the left. The tangential velocity always points in the direction of flow, except where $\sin\theta = 0$, i.e., for $\theta = 0$ and $\theta = \pm\pi$; these points are stagnation points. The tangential velocity is largest and equal to $2\underset{0}{v}$ when $\sin\theta$ has a maximum, i.e, for $\theta = \pm\pi/2$.

We apply the Bernoulli equation along the cylinder wall:

$$\frac{v^2}{2g} + \phi = \frac{2\underset{0}{v}^2}{g}\sin^2\theta + \phi = H_0. \tag{9.27}$$

The hydraulic head has a maximum for $\theta = \pm\pi$ and $\theta = 0$, where $|w|$ is smallest:

$$\phi_{\max} = \frac{p}{\rho g} + x_3 = H_0, \tag{9.28}$$

where x_3 is a constant. The minimum hydraulic head is at $\theta = \pi/2$ where $|w|$ is $2\underset{0}{v}$:

$$\phi_{\min} = \frac{p}{\rho g} + x_3 = H_0 - \frac{2\underset{0}{v}^2}{g}. \tag{9.29}$$

The streamlines for this case are shown in Figure 9.4 and the corresponding pressure distribution in Figure 9.5.

9.2.2 A Moving Cylinder

We determine the velocity and pressure fields for a cylinder that moves through a fluid that is at rest far away, by computing the velocity relative to moving coordinates attached to the object. The complex potential (9.18) corresponds to a stationary cylinder, with the velocity field given by $w = \underset{0}{w}$. This complex potential applies to the steady relative velocity field if the cylinder moves with complex velocity $-\underset{0}{w}$. We obtain an

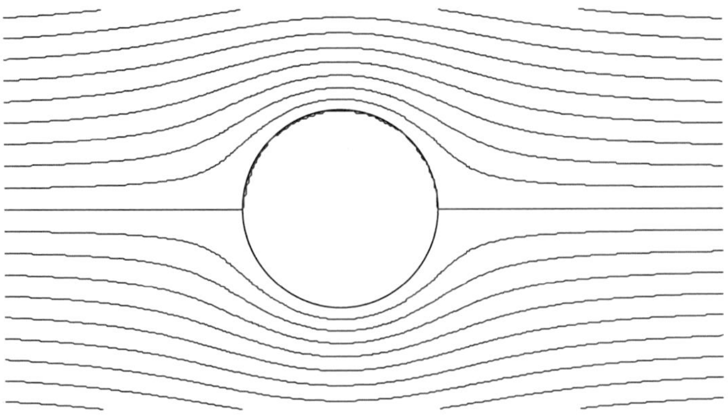

Figure 9.4: Flow with a uniform far field and a cylindrical impermeable object.

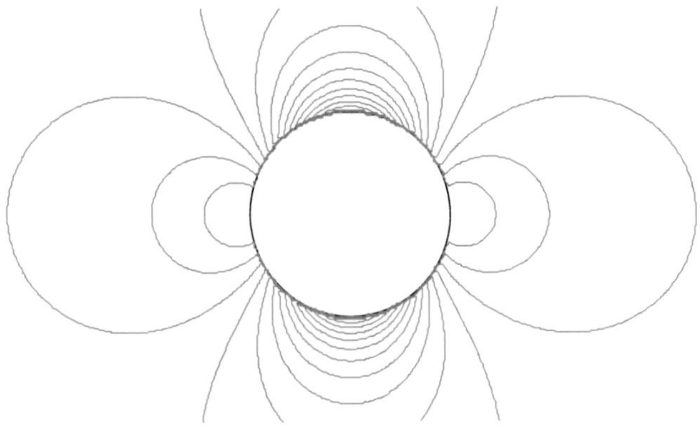

Figure 9.5: Contours of constant hydraulic head for the case of Figure 9.4.

expression for the global velocity field by adding a term $-\underset{0}{w}$ to the expression for w, which corresponds to adding a term $\underset{0}{w}z$ to the complex potential (9.18),

$$\Omega = -\left[\underset{0}{w}\underset{\sim}{z} + \underset{0}{\bar{w}}\frac{R^2}{\underset{\sim}{z}}\right] + \underset{0}{w}\underset{\sim}{z} = -\underset{0}{\bar{w}}\frac{R^2}{\underset{\sim}{z}}, \tag{9.30}$$

where $\underset{\sim}{z}$ is the local moving coordinate, with its origin at the center of the cylinder. We differentiate Ω with respect to z[1]

$$w = -\frac{d\Omega}{dz} = -\underset{0}{\bar{w}}\frac{R^2}{\underset{\sim}{z}^2}. \tag{9.31}$$

[1] z and $\underset{\sim}{z}$ differ only by a function of time; differentiation with respect to z and with respect to $\underset{\sim}{z}$ have the same result.

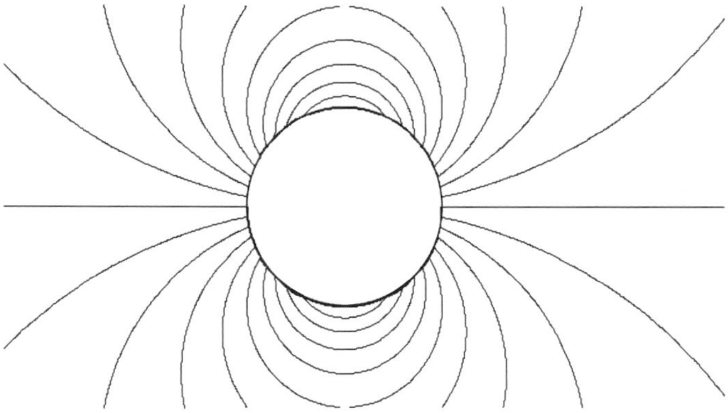

Figure 9.6: Instantaneous streamlines for the case of an object moving at constant speed through a fluid at rest far away

We consider the special case that $w_0 = v_0$, where v_0 is real and positive. We verify that the cylinder indeed moves with velocity $v_x = -v_0$ by setting $z = \pm R$, which indeed gives $w = -v_0$; at these two points the velocity of fluid and cylinder are equal. A contour plot of the imaginary part of (9.30), the stream function, is shown in Figure 9.6; the curves are the instantaneous streamlines, which indicate the direction of flow at any instant of time.

We obtain the expression for the velocity along the cylinder wall by adding a term $-v_0$ to (9.24),

$$ve^{-i\beta} = 2v_0 e^{-i(\theta-\pi/2)} \sin\theta - v_0 \tag{9.32}$$

The maximum values of the velocity on the cylinder wall remain at $\theta = \pm\pi/2$, but are reduced from $2v_0$ to v_0.

A Rotating Moving Cylinder

We next give the cylinder wall a rotation such that the fluid obtains an additional rotational component along the cylinder wall. This amounts to adding a vortex to the flow field, i.e., to adding a constant term $\overset{v}{v}_\theta$ to the tangential component of flow. This contribution adds to the value of v_θ, but, because $\sin\theta < 0$ for the lower half of the cylinder and $\sin\theta > 0$ for the upper half, the effect of the contribution differs for the upper and lower halves of the cylinder. If $\overset{v}{v}_\theta$ is negative, i.e., if the rotation is clockwise, then it reduces the velocity on the lower half of the cylinder, and increases it on the upper half. As a result, the pressure on the lower half will increase and on the upper half it will decrease; the cylinder will experience a resultant upward force. If the rotation is counter-clockwise, however, the resultant force will be down.

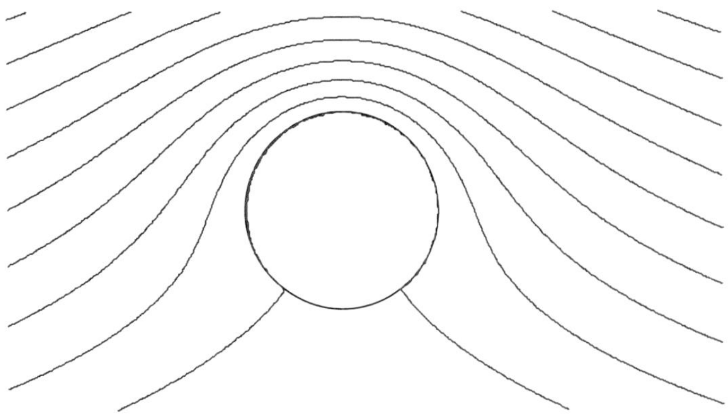

Figure 9.7: Flow with a uniform far field and a rotating cylindrical impermeable object.

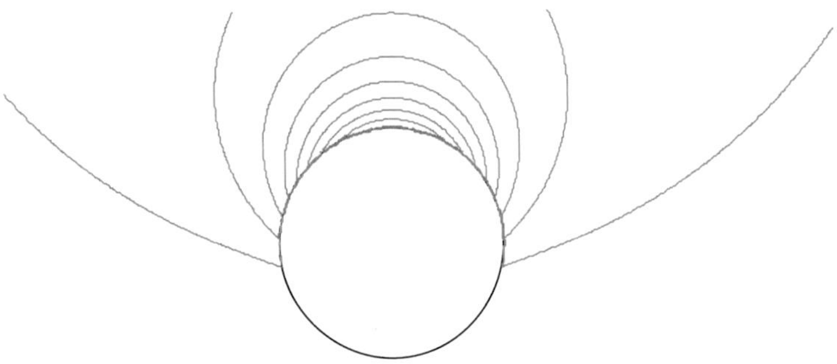

Figure 9.8: Contours of constant hydraulic head for the case of Figure 9.7

9.2.3 A Rotating Stationary Cylinder in a Moving Fluid

Figures 9.7 and 9.8 correspond to the case that the cylinder rotates, but remains in place. The presence of the rotational component $\overset{v}{v}_\theta$ causes the stagnation points to move from their original positions at $x_i = (\pm R, 0)$. The angles θ that correspond to the stagnation point locations are obtained by setting the velocity v_θ equal to zero. This yields for the case of flow in the positive x direction with velocity $\underset{0}{v}$

$$-2\underset{0}{v}\sin\theta + \overset{v}{v}_\theta = 0, \tag{9.33}$$

or

$$\sin\theta = \frac{\overset{v}{v}_\theta}{2\underset{0}{v}}. \tag{9.34}$$

A value for θ that satisfies this equation is possible only if the magnitude of the fraction to the right of the equal sign is less than 1. For the limiting case that $\overset{v}{v}_\theta = 2\underset{0}{v}$ we find $\theta = \pi/2$, and for $\overset{v}{v}_\theta = -2\underset{0}{v}$ we find $\theta = -\pi/2$. When $|\overset{v}{v}_\theta| > 2|\underset{0}{v}|$, the stagnation point

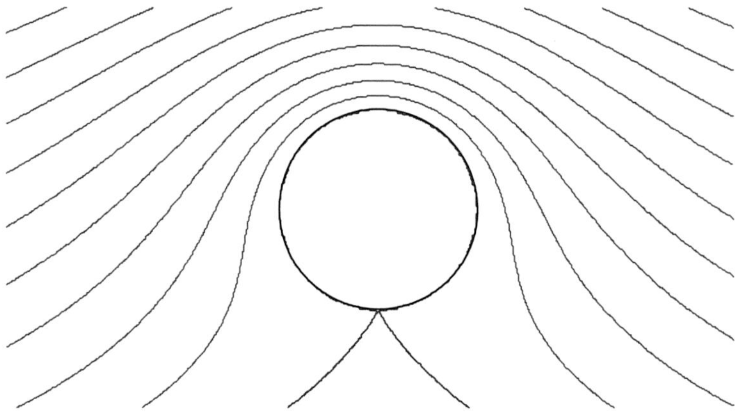

Figure 9.9: Streamlines for the case that there is a stagnation point at $\theta = -\frac{\pi}{2}$

lies in the fluid away from the cylinder wall. The path lines in Figure 9.9 correspond to the special case that the stagnation point is at $\theta = -\pi/2$.

9.2.4 Flow with Many Cylindrical Impermeable Objects

We consider the case of flow with an arbitrary number of impermeable inclusions, illustrated in Figure 9.10, see Strack [1989], Salisbury and Barnes [1994], Barnes and Janković [1999]. We apply a method called the analytic element method, which is based upon the superposition of a number of functions that possess degrees of freedom, capable of meeting the boundary conditions (see Strack [1989], Haitjema [1995], Strack [2003]). We first define an analytic element for a cylindrical impermeable object centered at $\underset{m}{z}$ and of radius $\underset{m}{R}$; the complex potential for this element, $\underset{m}{\Omega}$, is given by the following Laurent expansion of negative powers of a local dimensionless complex variable $\underset{m}{Z}$, defined as

$$\underset{m}{Z} = \frac{z - \underset{m}{z}}{\underset{m}{R}}. \tag{9.35}$$

The complex potential for the mth element is

$$\underset{m}{\Omega} = \sum_{n=1}^{\infty} \underset{m}{\alpha_n} \underset{m}{Z}^{-n}. \tag{9.36}$$

We construct the solution for N of these elements by adding one term for uniform flow to N complex potentials of the form (9.36)

$$\Omega = -\underset{0}{w} z + \sum_{m=1}^{N} \underset{m}{\Omega}(\underset{m}{Z}), \tag{9.37}$$

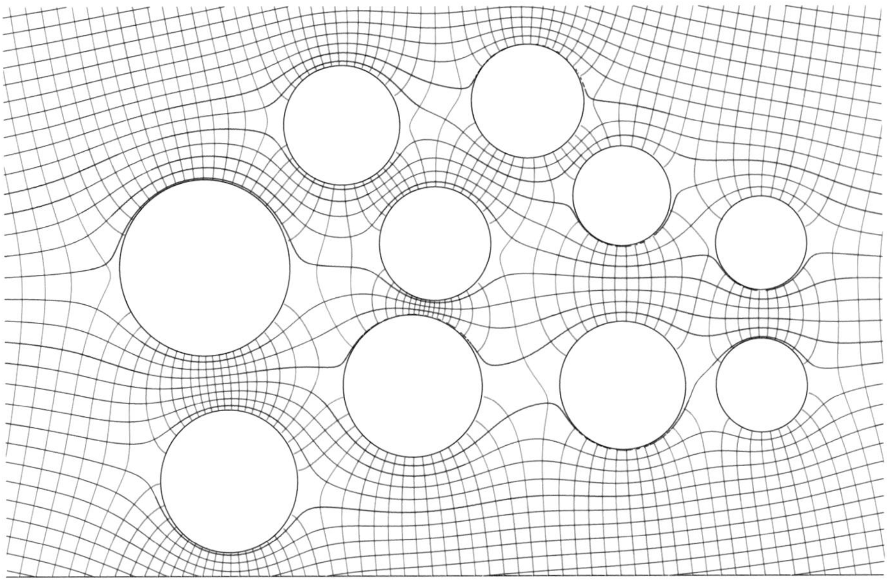

Figure 9.10: Ten impermeable cylinders in a field of uniform flow

where $\underset{0}{w}$ is a complex velocity, defined as

$$\underset{0}{w} = \underset{0}{v} e^{-i\alpha}, \tag{9.38}$$

and where $\underset{0}{v}$ is the magnitude of the velocity and α the angle between the velocity vector and the x-axis. We require that the cylinders do not intersect each other and determine the constants $\underset{m}{a_n}$ in such a way that the cylinder walls are impermeable, i.e., that $\Im\Omega = \Psi$ is constant along each of the cylinder walls.

Method of Solution

We solve the problem iteratively, i.e., we solve for the constants in the complex potential for one analytic element at a time, treating the constants in the complex potentials for all the other elements as knowns, following Barnes and Janković [1999]. The complex potential for any of the analytic elements is holomorphic outside the boundary associated with that element and thus can be expanded in a Taylor series about the center of the analytic element whose unknowns are solved for. We expand $\Omega(z)$ about the center of the cylinder, $\underset{j}{z}$,

$$\Omega(z) = \sum_{n=0}^{\infty} \underset{j}{a_n} \underset{j}{Z}^n \tag{9.39}$$

We compute the complex potential Ω using the current values of the constant parameters $\underset{m}{\alpha_n}$; initially these values are all zero. We obtain expressions for the complex constants $\underset{j}{a}_n$ by the use of (8.163) on page 108,

$$\underset{j}{a}_n = \frac{1}{2\pi \mathrm{i}} \oint\limits_{\mathscr{C}} \frac{\Omega(z(\delta))}{\delta^{n+1}} d\delta \qquad n = 0, 1, \cdots \qquad \delta \in \mathscr{C}, \tag{9.40}$$

where

$$\delta = \frac{z - \underset{j}{z}}{\underset{j}{R}} \qquad z(\delta) = \underset{j}{R}\delta + \underset{j}{z}. \tag{9.41}$$

We write $\delta = \mathrm{e}^{\mathrm{i}\theta}$ and use this in (9.40)

$$\underset{j}{a}_n = \frac{1}{2\pi \mathrm{i}} \oint\limits_{C} \Omega((z(\delta))\mathrm{e}^{-\mathrm{i}(n+1)\theta}\mathrm{e}^{\mathrm{i}\theta} d\mathrm{i}\theta = \frac{1}{2\pi} \oint\limits_{C} \Omega(z(\delta))\mathrm{e}^{-\mathrm{i}n\theta} d\theta. \tag{9.42}$$

Numerical evaluation of the integral is covered in Appendix A.

Solving for the Constants of Element j
The complex potential for element j is

$$\underset{j}{\Omega} = \sum_{n=1}^{\infty} \underset{j}{\alpha}_n \underset{j}{Z}^{-n}. \tag{9.43}$$

The boundary condition along the cylinder wall is that the stream function is constant. The complex potential has converged when it meets this condition with a prescribed precision. If not, the complex potential associated with cylinder j must be modified, i.e., the parameters $\underset{j}{\alpha}_n$ must be re-computed and the calculation cycle continued until the boundary condition is met with the required precision.

The complex potential (9.43) does not contribute to the contour integral (9.42) because the integral

$$\frac{1}{2\pi \mathrm{i}} \oint\limits_{\mathscr{C}} \frac{\underset{j}{\Omega}(z(\delta))}{\delta^{n+1}} d\delta = 0 \qquad n = 0, 1, \cdots \qquad \delta \in \mathscr{C} \tag{9.44}$$

vanishes; the integrand is holomorphic in the domain outside of $\mathscr{C}$, including infinity. This may be seen alternatively by substituting expression (9.43) for $\underset{j}{\Omega}$ in (9.44) and working out the integral.

We add the complex potential for element j to the Taylor series expansion about $\underset{j}{Z} = 0$ with its coefficients obtained from (9.42), and write the complex potential for points of circle j,

$$\Omega = \sum_{n=1}^{\infty} \underset{j}{a}_n \underset{j}{Z}^n + \sum_{n=1}^{\infty} \underset{j}{\alpha}_n \underset{j}{Z}^{-n} + \underset{j}{a}_0 \qquad \underset{j}{Z}\underset{j}{\bar{Z}} = 1. \tag{9.45}$$

The constant $\underset{j}{a}_0$ is added separately, so that both sums start at $n = 1$; the imaginary part of $\underset{j}{a}_0$ represents the constant value of the stream function along the cylinder wall. We use the condition that $\underset{j}{Z}$ represents a point on the boundary of the cylinder, i.e,

$$\underset{j}{Z}^{-n} = \underset{j}{\bar{Z}}^{n}, \tag{9.46}$$

so that (9.45) becomes

$$\Omega = \sum_{n=1}^{\infty} \underset{j}{a_n} \underset{j}{Z}^{n} + \sum_{n=1}^{\infty} \underset{j}{\alpha_n} \underset{j}{\bar{Z}}^{n} + \underset{j}{a_0} \qquad \underset{j}{Z}\underset{j}{\bar{Z}} = 1. \tag{9.47}$$

The boundary condition is that the stream function Ψ be constant along the cylinder wall; this condition is met if the sum of the two first terms in (9.47) is real, i.e., if the constants $\underset{j}{\alpha_n}$ satisfy the following conditions for all n:

$$\underset{j}{\alpha_n} = \underset{j}{\bar{a}_n} \qquad n > 0 \qquad j = 1, 2, \ldots, N, \tag{9.48}$$

so that the complex potential (9.47) has a constant imaginary part, equal to the imaginary part of the constant $\underset{j}{a_0}$.

The iterative process consists of expanding the sum of all contributions to the complex potential about the center of each cylinder in turn and then updating the coefficients of the complex potential representing the cylinder under consideration.

9.2.5 Two-Dimensional Horizontal Flow toward a Sink

We consider the case of two-dimensional flow to a point sink of discharge Q. The outer boundary of the domain is a circle of radius R and the sink is at the origin. The condition along the circular boundary is that there is a clockwise tangential component of flow of magnitude $\underset{0}{v}$. The solution to this problem consists of the sum of a vortex and a point sink. The complex potential Ω is the sum of (9.7) and (9.12),

$$\Omega = \frac{Q}{2\pi} \ln \frac{z}{R} + \frac{A}{2\pi \mathrm{i}} \ln \frac{z}{R}. \tag{9.49}$$

where we added a constant $-\frac{A}{2\pi \mathrm{i}} \ln R$ to the term that represents the vortex, to obtain a dimensionless argument; this merely adds a constant to the stream function, which is immaterial; only differences in value of the stream function matter. The complex velocity function is

$$w = -\frac{Q - \mathrm{i}A}{2\pi} \frac{1}{z}. \tag{9.50}$$

The vortex creates a tangential component of flow, and the sink a radial one. The tangential component along $z\bar{z} = R^2$ points in the direction of $\theta - \pi/2$, so that the contribution to w of the vortex, $\underset{v}{w}$ is equal to $\underset{0}{v} e^{-\mathrm{i}\theta + \mathrm{i}\pi/2}$,

$$\underset{v}{w} = \underset{0}{v} e^{-\mathrm{i}\theta + \mathrm{i}\pi/2} = \frac{\mathrm{i}A}{2\pi} \frac{1}{R} e^{-\mathrm{i}\theta}, \tag{9.51}$$

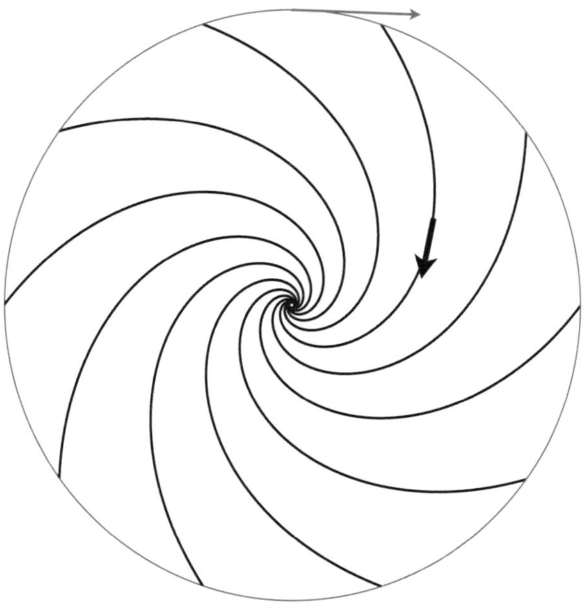

Figure 9.11: A combination of a point sink and a vortex.

so that

$$A = 2\pi v_0 R. \tag{9.52}$$

The expression for the complex velocity, (9.50), is

$$w = -\frac{Q - 2\pi \mathrm{i} v_0 R}{2\pi}\frac{1}{z}. \tag{9.53}$$

The magnitude of the square of the velocity vector at a point z is:

$$v^2 = \frac{Q^2 + 4\pi^2 v_0^2 R^2}{4\pi^2}\frac{1}{z\bar{z}} = \frac{Q^2 + (2\pi v_0 R)^2}{(2\pi r)^2}. \tag{9.54}$$

The velocity is constant for constant values of $r = \sqrt{z\bar{z}}$; we obtain an expression for the hydraulic head from the Bernoulli equation, which holds throughout the flow domain with the exception of the origin, i.e.,

$$\phi = \frac{p}{\rho g} + x_3 = H_0 - \frac{v^2}{2g}, \tag{9.55}$$

with v^2 given by (9.54). The contour lines of the hydraulic head are circles; the hydraulic head decreases as $-1/r^2$ toward the point sink and is constant along circles centered at the point sink. The value of the constant, H_0, is obtained from the values of ϕ and v at the outer boundary $r = R$.

We obtain equations for the streamlines by setting the stream function, the imaginary part of (9.49), equal to constant values. We obtain an expression for Ψ from (9.49),

$$\Psi = \frac{Q}{2\pi}\theta - \frac{A}{2\pi}\ln\frac{|z|}{R} = \frac{Q}{2\pi}\theta - v_0 R\ln\frac{r}{R}. \tag{9.56}$$

We set Ψ equal to some constant, Ψ_0, and solve for r/R:

$$\frac{r}{R} = \exp\left[\frac{1}{v_0 R}\left(\frac{Q}{2\pi}\theta - \Psi_0\right)\right], \tag{9.57}$$

where Ψ_0 is the constant value of the stream function along the streamline. The curves defined by (9.57) are logarithmic spirals; points of the curves may be determined by letting θ take on values between $-\pi$ and π and compute the corresponding values of r.

A plot of the streamlines is shown in Figure 9.11; note the jump in the stream function resulting from the net flow into the point sink. This is due to the jump in the stream function along the negative x-axis, where θ jumps from π to $-\pi$, which corresponds to a discontinuity of Q in the stream function.

9.2.6 Rotational Flow with a Forced Vortex

The case of a forced vortex differs from the problems considered thus far in that the curl of the velocity field is not zero, but equal to some constant value C. The basic equation that relates the vector field to the curl and divergence, (8.92), on page 96, becomes

$$2\frac{\partial w}{\partial \bar{z}} = -\mathrm{i}\delta = -\mathrm{i}C. \tag{9.58}$$

Integration gives

$$w = f(z) - \mathrm{i}\tfrac{1}{2}C\bar{z}. \tag{9.59}$$

We consider a case of purely rotational flow, with a forced vortex in the domain $0 \leq r \leq R$. The function $f(z)$ reduces to zero,

$$w = v\mathrm{e}^{-\mathrm{i}\alpha} = -\mathrm{i}\tfrac{1}{2}C\bar{z} = -\mathrm{i}\tfrac{1}{2}Cr\mathrm{e}^{-\mathrm{i}\theta} = \tfrac{1}{2}Cr\mathrm{e}^{-\mathrm{i}(\theta+\pi/2)}. \tag{9.60}$$

The velocity is perpendicular to the radial coordinate and points in the direction of increasing θ in an (r, θ)-coordinate system, provided that C is positive. If the tangential velocity component is v_0 for $r = R$, then

$$C = v_0\frac{2}{R}, \tag{9.61}$$

so that the expression for w is:

$$w = -\mathrm{i}v_0\frac{\bar{z}}{R}. \tag{9.62}$$

The flow is rotational and therefore the energy head H is not constant. We obtain an expression for $\nabla^2 H$ from (3.46) on page 25. The third term to the right of the equal sign in (3.46) is zero because the velocity is a linear function of the coordinates. Furthermore, $C^2 = 4v_0^2/R^2$, so that (3.46) becomes:

$$\nabla^2 H = 4\frac{\partial^2 H}{\partial z \partial \bar{z}} = \frac{4v_0^2}{gR^2} \qquad 0 \le r \le R. \tag{9.63}$$

We integrate this first with respect to z and then with respect to $\bar{z}$,

$$H = \frac{v_0^2}{g}\frac{z\bar{z}}{R^2} + f(z) + \overline{f(z)} \qquad 0 \le r \le R, \tag{9.64}$$

where $f(z)$ is an arbitrary holomorphic function. The flow outside the forced vortex is irrotational (this irrotational flow is called a free vortex) represented by the complex potential

$$\Omega = \frac{A}{2\pi \mathrm{i}} \ln \frac{z}{R} + C_1, \tag{9.65}$$

where C_1 is a constant. The complex velocity vector is

$$w = -\frac{d\Omega}{dz} = -\frac{A}{2\pi \mathrm{i}}\frac{1}{z}. \tag{9.66}$$

We obtain the complex velocity at $r = R$ from (9.62):

$$w = v_0\frac{R}{R}\mathrm{e}^{-\mathrm{i}(\theta+\pi/2)}. \tag{9.67}$$

The velocity is continuous and the latter equation must match (9.66) for $z = R\mathrm{e}^{\mathrm{i}\theta}$, so that, with $1/\mathrm{i} = \mathrm{e}^{-\mathrm{i}\pi/2}$

$$-\frac{A}{2\pi}\frac{1}{R}\mathrm{e}^{-\mathrm{i}(\theta+\pi/2)} = v_0\mathrm{e}^{-\mathrm{i}(\theta+\pi/2)}. \tag{9.68}$$

We solve this for A:

$$A = -2\pi R v_0. \tag{9.69}$$

The expression for the velocity outside the forced vortex area becomes

$$w = v_0\frac{R}{z}\mathrm{e}^{-\mathrm{i}\pi/2} = v_0\frac{R}{r}\mathrm{e}^{-\mathrm{i}(\theta+\pi/2)} \qquad R \le r. \tag{9.70}$$

The flow is irrotational in the free vortex area, so that H is equal to a constant there, i.e.,

$$H = B \qquad R \le r. \tag{9.71}$$

The value of this constant equals the value of H, see (3.49), at $r = R$. Since the energy head is constant in $r \ge R$, we can compute B from the value of H at infinity. Since the velocity approaches zero for $r \to \infty$, we obtain

$$B = \phi_0, \tag{9.72}$$

where $\phi_0 = p/(\rho g) + x_3$ is the hydraulic head at infinity. The expression for H valid inside the circle of radius R, given by (9.64) must match the constant value ϕ_0 for $r = R$, so that

$$\frac{v_0^2}{g} + f(z) + f(\bar{z}) = \phi_0 \qquad z\bar{z} = R^2. \tag{9.73}$$

The function $f(z)$ is holomorphic for $z\bar{z} \leq R^2$ and is constant along the boundary; this function must be a constant everywhere,

$$f(z) + \overline{f(z)} = \phi_0 - \frac{v_0^2}{g}. \tag{9.74}$$

The expression for the energy head, (9.64), inside the circle of radius R reduces to

$$H = \frac{v_0^2}{g}\frac{z\bar{z} - R^2}{R^2} + \phi_0 \qquad z\bar{z} \leq R^2. \tag{9.75}$$

We obtain an expression for the hydraulic head in the forced vortex area by subtracting $v^2/(2g)$ from H. We find from (9.62) that v equals $v_0 r/R$ in the area of the forced vortex so that

$$\begin{aligned}\phi = \frac{p}{\rho g} + x_3 = H - \frac{v_0^2}{2g}\left(\frac{r}{R}\right)^2 &= \frac{v_0^2}{2g}\frac{2r^2 - r^2 - 2R^2}{R^2} + \phi_0 \\ = \phi_0 - \frac{v_0^2}{2g}\frac{2R^2 - r^2}{R^2} &\qquad 0 \leq r \leq R.\end{aligned} \tag{9.76}$$

We obtain an expression for the hydraulic head in the free vortex area in a similar manner

$$\phi = \phi_0 - \frac{v_0^2}{2g}\left(\frac{R}{r}\right)^2 \qquad R \leq r \leq \infty. \tag{9.77}$$

Thus, the hydraulic head increases from the value $H_0 = \phi_0 - v_0^2/g$ at $r = 0$ quadratically with r until $r = R$, where it reaches the value $\phi_0 - v_0^2/(2g)$; afterward, the hydraulic head increases with r hyperbolically to a limiting value of ϕ_0 at infinity. The increase in hydraulic head between $r = 0$ and $r = R$ is exactly equal to the increase between $r = R$ and infinity.

The Stream Function

We can use a stream function because the flow is divergence free, and contour it to obtain a plot of streamlines. The velocity in the forced vortex area is given by (9.62) and we obtain the stream function upon integration of (8.100) on page 97,

$$w = -2\mathrm{i}\frac{\partial \Psi}{\partial z} = -\mathrm{i}v_0\frac{\bar{z}}{R}, \tag{9.78}$$

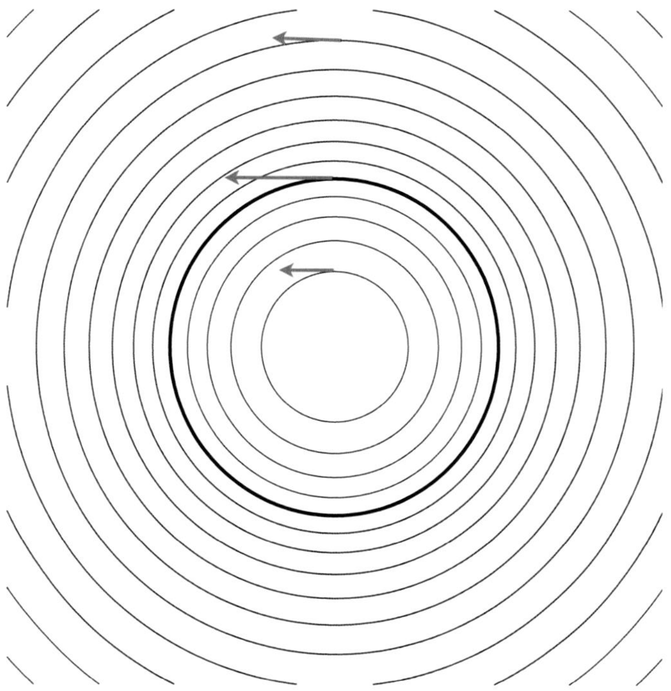

Figure 9.12: Streamlines for forced and free vortices.

so that

$$\Psi = \tfrac{1}{2} v_0 \frac{z\bar{z}}{R}. \tag{9.79}$$

The stream function in the free vortex area is the imaginary part of the complex potential valid for $r \geq R$, (9.65), with A from (9.69)

$$\Psi = -\frac{A}{2\pi} \ln r + C_1 = R v_0 \ln \frac{r}{R} + C_1. \tag{9.80}$$

We choose the constant C_1 such that the stream function is continuous at $r = R$; $C_1 = \tfrac{1}{2} v_0 R$, i.e.,

$$\Psi = R v_0 \ln \frac{r}{R} + \tfrac{1}{2} v_0 R. \tag{9.81}$$

Contours of the stream function are shown in Figure 9.12. A plot is shown in Figure 9.13 of the dimensionless energy head, the velocity head, and the hydraulic head (in this case equal to the pressure head if we set $x_3 = 0$), for both the domains inside the forced vortex ($0 \leq r/R \leq 1$) and outside ($r/R \leq 2$).

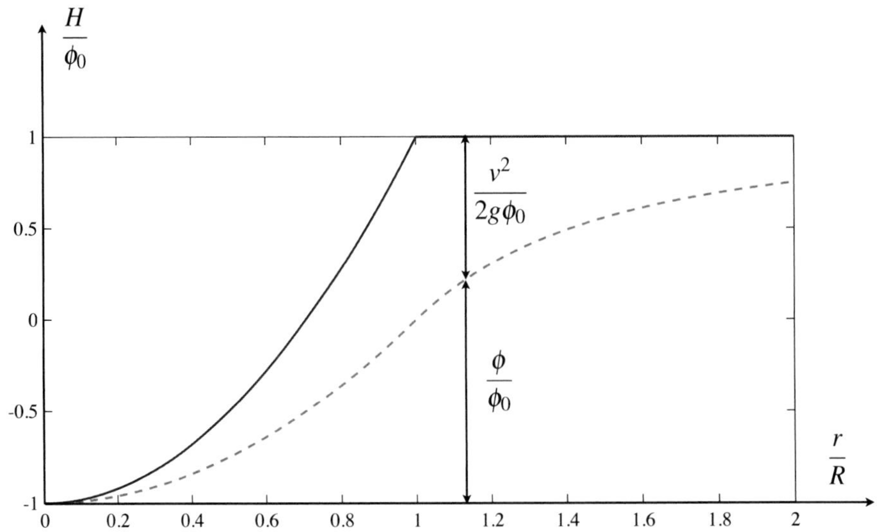

Figure 9.13: Dimnesionless energy head, H/ϕ_0, hydraulic head, ϕ/ϕ_0, and velocity head $v^2/(2g\phi_0)$ for forced and free vortices ($x_3 = 0$) versus dimensionless radial distance, r/R

9.3 Conformal Mapping

Any function of a single complex variable, z, is a holomorphic function, with the exception of singular points. A technique for solving boundary-value problems based on this principle is called *conformal mapping*, and consists of transforming the boundary of the domain, in the case of our example an ellipse, into a simpler boundary shape, in our example, a circle. The transformation, or mapping, of the domain is called conformal, because orthogonality of lines is preserved, and the scaling factors in two orthogonal directions are the same. We see this in Figure 9.3 on page 113, where a network of equipotentials and streamlines is displayed. The spaces between the intersecting blue and red curves look more and more like squares as we zoom in, i.e., increasingly reduce the spaces between intersecting curves. Tables exist of conformal transformations, that can be used to select one that fits one particular boundary-value problem. We apply conformal mapping to solve the problem of an impermeable elliptical object in a field of uniform flow, which can be generalized to the case of an ellipse moving through an incompressible in-viscid fluid at rest far away.

9.3.1 Transforming an Ellipse into a Circle

We solve the problem of an elliptical impermeable object in a field of uniform flow. The boundary conditions are that far away the flow is in the x-direction, and that the boundary of the ellipse centered at the origin and with foci at z_1 and z_2, is impermeable. We transform the ellipse first into an ellipse with foci at $Z = -1$ and $Z = 1$ in a dimensionless Z-plane, defined by the transformation:

$$Z = \frac{z - \frac{1}{2}(z_1 + z_2)}{\frac{1}{2}(z_2 - z_1)}, \tag{9.82}$$

with the reverse transformation:

$$z = \tfrac{1}{2}(z_2 - z_1)Z + \tfrac{1}{2}(z_1 + z_2). \tag{9.83}$$

We next transform the ellipse into a circle in the plane of another dimensionless parameter, χ, as shown in Figure 9.14, where the red slot in the z-plane that connects the foci of the ellipse corresponds to the circle of radius ν in the χ-plane, and the ellipse itself to the unit circle. The advantage of this transformation is that we have the solution to the problem of a cylindrical object, such as that in Figure 9.14c, in a field of uniform flow, and that the functions $\chi(z)$ and $z(\chi)$ make it possible to apply that solution in a physical plane, the z-plane. The function that performs this transformation is:

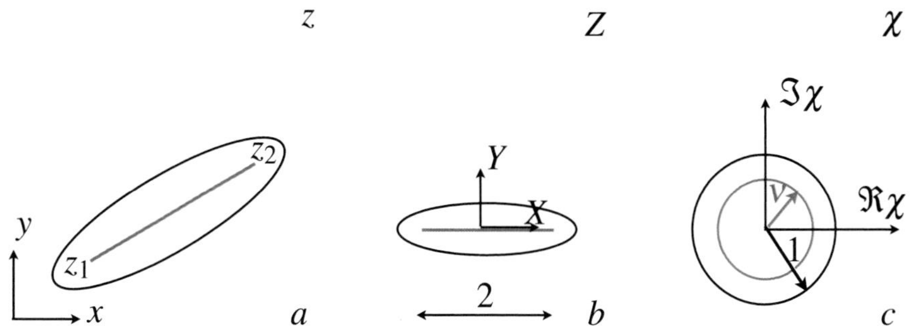

Figure 9.14: The domains in the z, Z, and χ-planes

$$Z = \frac{1}{2}\left[\frac{\chi}{\nu} + \frac{\nu}{\chi}\right]. \tag{9.84}$$

We verify that points on the unit circle in the χ-plane indeed correspond to the ellipse in the Z-plane; we set $\chi = \mathrm{e}^{\mathrm{i}\theta}$ in (9.84):

$$Z = \frac{1}{2}\left[\frac{1}{\nu}\mathrm{e}^{\mathrm{i}\theta} + \nu\mathrm{e}^{-\mathrm{i}\theta}\right] = \frac{1}{2}\frac{1}{\nu}(\cos\theta + \mathrm{i}\sin\theta) + \frac{1}{2}\nu(\cos\theta - \mathrm{i}\sin\theta), \tag{9.85}$$

or

$$Z = \frac{1}{2}\left(\frac{1}{\nu} + \nu\right)\cos\theta + \mathrm{i}\frac{1}{2}\left(\frac{1}{\nu} - \nu\right)\sin\theta. \tag{9.86}$$

We separate the real and imaginary parts:

$$\begin{aligned} X &= \frac{1}{2}\left(\frac{1}{\nu} + \nu\right)\cos\theta \\ Y &= \frac{1}{2}\left(\frac{1}{\nu} - \nu\right)\sin\theta, \end{aligned} \tag{9.87}$$

so that

$$\left(\frac{X}{A}\right)^2+\left(\frac{Y}{B}\right)^2=1. \tag{9.88}$$

The half-lengths of the principal axes of the ellipse in the dimensionless Z-plane are:

$$\begin{aligned} A &= \frac{1}{2}\left(\frac{1}{\nu}+\nu\right) \\ B &= \frac{1}{2}\left(\frac{1}{\nu}-\nu\right). \end{aligned} \tag{9.89}$$

We express ν in terms of the ratio of the minor to major principal axes, B/A. We use (9.89):

$$\frac{B}{A}=\frac{1-\nu^2}{1+\nu^2}, \tag{9.90}$$

so that:

$$\nu=\sqrt{\frac{1-B/A}{1+B/A}}. \tag{9.91}$$

If the magnitudes of the major axes, a and b, are given, we obtain the distance between the two focal points, L, from either:

$$L=2\frac{a}{A}, \tag{9.92}$$

or from

$$L=2\frac{b}{B}, \tag{9.93}$$

where A and B are given by (9.89), and L is defined by

$$z_2-z_1=Le^{\mathrm{i}\alpha}. \tag{9.94}$$

The transformation (9.84) returns Z for any given value of χ; we obtain values of χ for given values of Z by reversing the transformation:

$$\left(\frac{\chi}{\nu}\right)^2-2Z\frac{\chi}{\nu}+1=0, \tag{9.95}$$

or

$$\frac{\chi}{\nu}=Z+\sqrt{(Z-1)(Z+1)}. \tag{9.96}$$

The plus sign in front of the square root applies because the exterior of the ellipse corresponds to the exterior of the circle; χ approaches infinity when Z approaches infinity. Care must be taken in evaluating the square root in this transformation; it is necessary to evaluate the function by computing the square roots of $Z-1$ and $Z+1$ separately, and then compute the product. We see this by considering the points $Z=\mathrm{i}$ and $Z=-\mathrm{i}$. If we compute the square root of Z^2-1, we obtain $\sqrt{-2}$ for both points, which is incorrect; we expect these points to be mapped from the Z-plane onto the χ-plane as shown in Figure 9.15. We achieve this by computing the square root as

$\sqrt{Z-1}\sqrt{Z+1}$. For the point $Z=\mathrm{i}$, the angle between the X-axis and the line that connects $Z=1$ to $Z=\mathrm{i}$ is $3\pi/4$, and the length of this line is $\sqrt{2}$. The angle between the X-axis and a line that connects $Z=-1$ to $Z=\mathrm{i}$ is $\pi/4$. Thus the complex numbers $Z-1$ and $Z+1$ can be written for $Z=\mathrm{i}$ as:

$$(\mathrm{i}-1)=\sqrt{2}\mathrm{e}^{3\pi\mathrm{i}/4} \qquad (\mathrm{i}+1)=\sqrt{2}\mathrm{e}^{\pi\mathrm{i}/4}, \tag{9.97}$$

so that

$$\sqrt{\mathrm{i}-1}\sqrt{\mathrm{i}+1}=\sqrt{2}\mathrm{e}^{3\pi\mathrm{i}/8}\mathrm{e}^{\pi\mathrm{i}/8}=\sqrt{2}\mathrm{e}^{\mathrm{i}\pi/2}=\sqrt{2}\mathrm{i}. \tag{9.98}$$

This corresponds to points $\mathscr{A}^{+}$ in the Z- and χ-planes shown in Figure 9.15. If we

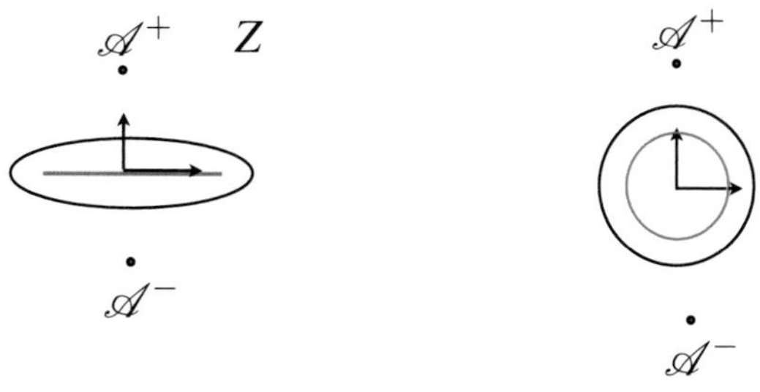

Figure 9.15: Mapping of points using the square root function

consider point $\mathscr{A}^{-}$, the angles change from $3\pi/4$ and $\pi/4$ to $-3\pi/4$ and $-\pi/4$, i.e., the signs are reversed, resulting in the locations of the points $\mathscr{A}^{-}$, shown in Figure 9.15.

9.3.2 An Elliptical Impermeable Object in Uniform Flow

With the ellipse transformed into a circle, we can solve the problem of an elliptical impermeable object in a field of uniform flow. We solve the problem in the χ-plane in the same way as we solved the problem of a cylindrical impermeable object in a field of uniform flow, but we must adapt the uniform flow field, as it will be transformed as a result of the conformal mapping. We consider that the uniform flow is in the x-direction and of magnitude $w=\underset{0}{v}$. Far away from the object, the velocity approaches the uniform value of $\underset{0}{v}$, and we find the corresponding expression for the complex potential from $\Omega=-\int w dz$, i.e.,

$$\Omega=-\underset{0}{v}z+C, \tag{9.99}$$

where C is an arbitrary constant. We find the corresponding expression in terms of χ after first transforming z into Z, using (9.83)

$$\Omega=-\tfrac{1}{2}\underset{0}{v}(z_2-z_1)Z-\tfrac{1}{2}\underset{0}{v}(z_1+z_2)+C. \tag{9.100}$$

We obtain the expression for the far-field in the χ-plane, i.e., the form of the complex potential for large values of χ, by replacing Z by $\frac{1}{2}(\chi/\nu+\nu/\chi)$, leaving out the term

ν/χ as it vanishes for $\chi \to \infty$, and leaving out the constant as it can be absorbed into a general constant afterward:

$$\Omega = -\tfrac{1}{4}\nu_0(z_2 - z_1)\frac{\chi}{\nu} + C^*, \tag{9.101}$$

where C^* is a new constant. We write $z_2 - z_1$ in terms of polar coordinates according to (9.94):

$$z_2 - z_1 = Le^{i\alpha}, \tag{9.102}$$

where L is the distance between the foci of the ellipse and α the angle between the orientation of the major principal axis and the x-axis. We obtain an expression for the complex far-field velocity in the χ-plane from (9.102):

$$w^* = -\frac{d\Omega}{d\chi} = \frac{1}{4}\nu_0(z_2 - z_1)\frac{1}{\nu} = \tfrac{1}{4}\nu_0\frac{Le^{i\alpha}}{\nu}. \tag{9.103}$$

We obtain the solution for the flow in the χ-plane from (9.18) on page 114:

$$\Omega = -\left[w_0^*\chi + \bar{w}_0^*\frac{1}{\chi}\right], \tag{9.104}$$

where we have chosen a new constant such that it cancels C^*. A flow net of equipotentials and streamlines for an ellipse in a field of uniform flow at an angle of 30° to the x-axis is shown in Figure 9.16.

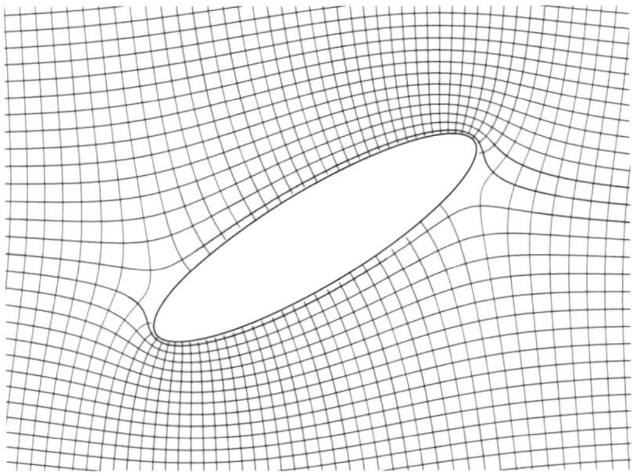

Figure 9.16: An ellipse in a field of uniform flow in the x-direction. The ratio b/a is .25

9.4 Groundwater flow

The basic equation for groundwater flow is Darcy's law:

$$q_i = -k\frac{\partial\phi}{\partial x_i}, \tag{9.105}$$

where q_i is the specific discharge vector [L/T]. The magnitude of this vector equals the amount of water flowing through a unit area normal to the direction of flow. The proportionality constant k [L/T] in (9.105) is the hydraulic conductivity, and ϕ is the hydraulic head [L], equal to the elevation above some reference level to which groundwater rises in a standpipe. A standpipe is a pipe, usually with a short screen at the bottom, drilled into the saturated zone of an aquifer.

For steady groundwater flow, the divergence of the specific discharge vector is zero:

$$\frac{\partial q_i}{\partial x_i} = 0 \tag{9.106}$$

The flow is irrotational if the hydraulic conductivity is isotropic, as is the case of (9.105), and piecewise constant, and if the density and viscosity of the fluid are constant. In that case a specific discharge potential φ exists such that q_i is minus its gradient,

$$q_i = -\partial_i(k\phi) = -\partial_i\varphi. \tag{9.107}$$

Since the flow is irrotational and divergence-free for this case, solutions are obtained as for fluid mechanics. The difference is that the term $v^2/(2g)$ is small in the case of groundwater flow, whereas the energy loss is substantial, caused by the resistance to flow by the porous medium. The energy head is approximately equal to the hydraulic head; the potential has a direct relationship to an observable quantity, the hydraulic head, as opposed to the case of open-channel flow.

9.4.1 Horizontal Confined Flow

We consider the case of horizontal flow in an aquifer bounded above and below by horizontal impermeable boundaries. The water bearing layer, or aquifer, is the permeable material in between the boundaries. If the thickness of this aquifer is H, then the vertically integrated flow over the thickness of the aquifer is Q_i, with

$$Q_i = -\frac{\partial kH\phi}{\partial x_i} = \frac{\partial \Phi}{\partial x_i}, \tag{9.108}$$

where Φ is the discharge potential for horizontal confined flow. The continuity equation (9.106) leads to

$$\frac{\partial Q_i}{\partial x_i} = 0, \tag{9.109}$$

so that the discharge potential is harmonic,

$$\nabla^2\Phi = 0. \tag{9.110}$$

Groundwater flow in a horizontal confined aquifer can be described by a complex potential $\Omega = \Phi + \mathrm{i}\Psi$. We define the complex discharge function W as

$$W = Q_x - \mathrm{i}Q_y, \tag{9.111}$$

and Darcy's law in complex form becomes,

$$W = -\frac{d\Omega}{dz}. \tag{9.112}$$

A fundamental difference between groundwater flow and open channel flow, is that the discharge potential is proportional to the hydraulic head, which has direct physical meaning. By contrast, the velocity potential in open channel flow does not; only its derivative has physical meaning. A vortex can exist in groundwater flow only on the boundary, since the interior jump in discharge potential that would exist in the case of a vortex inside the domain, is physically impossible. We illustrate this in the following example.

Uniform Flow Toward a Lock in a River

Consider the flow in a confined aquifer adjacent to a river with a lock at $x = 0$, $y = 0$, shown in Figure 9.17. The aquifer is the half plane $y \geq 0$, and the boundary conditions for the head ϕ are

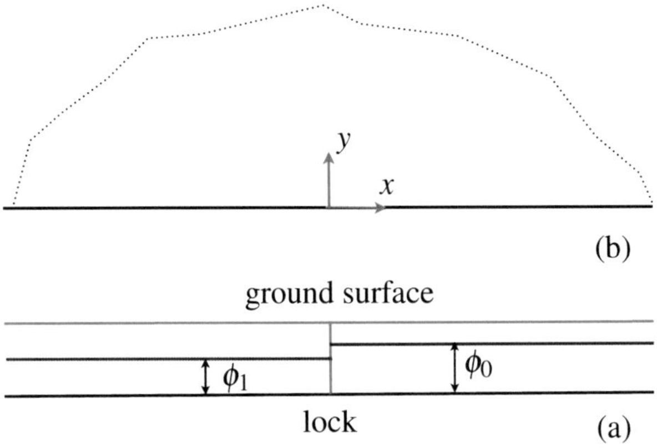

Figure 9.17: A lock in a river, with sectional view (a), and plan view (b)

$$\begin{aligned} x > 0,\ y = 0 \qquad & \phi = \phi_0 \qquad \Phi = \Phi_0 = kH\phi_0 \\ x < 0,\ y = 0 \qquad & \phi = \phi_1 \qquad \Phi = \Phi_1 = kH\phi_1 \end{aligned} \tag{9.113}$$

There is uniform flow with discharge $-Q_{y0}$ toward the river ($Q_{y0} < 0$). The complex potential with a real part that meets the boundary conditions is

$$\Omega = \mathrm{i}Q_{y0}z + \frac{\Phi_1 - \Phi_0}{\pi \mathrm{i}} \ln z + \Phi_0 \tag{9.114}$$

A flow net applicable to this case is shown in Figure 9.18, valid for the case that $\phi_1 = 20$ m, $\phi_0 = 18$ m, $k = 10$ m/d, $H = 10$ m, and $Q_{y0} = 2$ m^2/d.

Problem 9.1

Questions:

1. Demonstrate that (9.114) meets the boundary conditions.

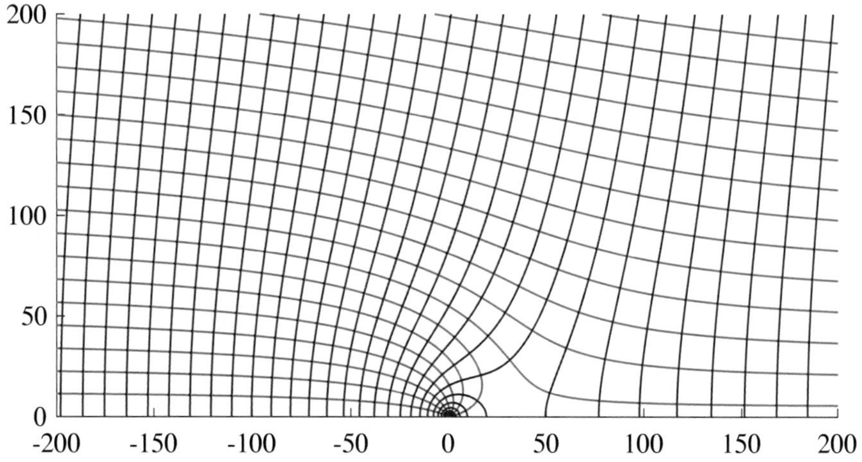

Figure 9.18: Flow net with contours of constant Φ (in red) and constant Ψ, in blue

9.5 Linear Elasticity

Problems involving linear elasticity are governed by Hooke's law, which relates the strain tensor e_{ij} to the stress tensor σ_{ij} in combination with the equilibrium equations. We consider cases of plane strain; there are no strains normal to the (x,y)-plane. We take compressive stresses as positive, as is usual for geological materials such as soil and rock.

There are numerous textbooks on linear elasticity; the texts Sokolnikoff [1956] and Green and Zerna [1968] are relatively close to the presentation given here, but go into considerably more depth.

9.5.1 Basic Equations

The basic equations are Hooke's law and the equilibrium equations. We begin by rewriting Hooke's law in terms of complex variables, and then do the same for the equilibrium equations.

Hooke's Law

Hooke's law for the case of plane strain is:

$$\sigma_{11} - \sigma_{22} = 2G(e_{11} - e_{22}) \tag{9.115}$$

$$\sigma_{12} = 2Ge_{12} \tag{9.116}$$

$$\sigma_{kk} = \frac{2G}{1-2\nu} e_{kk}, \tag{9.117}$$

where G [F/L^2] is the shear modulus and ν [.] is Poisson's ratio. Equation (9.117) is valid only if $\nu \neq 1/2$, i.e., if the material is not incompressible, a case that requires special treatment and we will not consider.

We introduce the complex displacement $w = u_x - \mathrm{i}u_y$ where u_x and u_y are the components of the displacement vector:

$$w = u_x - \mathrm{i}u_y. \tag{9.118}$$

We represent the transformed stresses and strains as τ^{mn} and S^{mn}, respectively:

$$\tau^{mn} = \tilde{\sigma}^{mn}, \tag{9.119}$$

and

$$S^{mn} = \tilde{e}^{mn}. \tag{9.120}$$

We transform the displacement gradient tensor α_{mn} into a purely covariant tensor $\tilde{\alpha}^{mn}$,

$$\tilde{\alpha}^{mn} = \overset{m}{\gamma}_i \overset{n}{\gamma}_j \alpha_{ij}. \tag{9.121}$$

The covariant base vectors are given by (8.42) and (8.43)

$$\overset{1}{\gamma}_i = \frac{\partial \tilde{x}_1}{\partial x_i} = \left(\frac{\partial \bar{z}}{\partial x}, \frac{\partial \bar{z}}{\partial y}\right) = (1, -\mathrm{i}) \tag{9.122}$$

$$\overset{2}{\gamma}_i = \frac{\partial \tilde{x}_2}{\partial x_i} = \left(\frac{\partial z}{\partial x}, \frac{\partial z}{\partial y}\right) = (1, \mathrm{i}). \tag{9.123}$$

We need to obtain expressions for two of the components, $\tilde{\alpha}^{11}$ and $\tilde{\alpha}^{12}$ only; the other two are the complex conjugates of the first two. We obtain for $\tilde{\alpha}^{11}$:

$$\begin{aligned}\tilde{\alpha}^{11} &= \tilde{\partial}^1 \tilde{u}^1 = \overset{1}{\gamma}_i \overset{1}{\gamma}_j \alpha_{ij} = \overset{1}{\gamma}_1 \overset{1}{\gamma}_1 \alpha_{11} + \overset{1}{\gamma}_2 \overset{1}{\gamma}_2 \alpha_{22} + \overset{1}{\gamma}_1 \overset{1}{\gamma}_2 \alpha_{12} + \overset{1}{\gamma}_2 \overset{1}{\gamma}_1 \alpha_{21} \\ &= \alpha_{11} - \alpha_{22} - \mathrm{i}(\alpha_{12} + \alpha_{21}),\end{aligned} \tag{9.124}$$

and for $\tilde{\alpha}^{12}$:

$$\begin{aligned}\tilde{\alpha}^{12} &= \tilde{\partial}^1 \tilde{u}^2 = \overset{1}{\gamma}_i \overset{2}{\gamma}_j \alpha_{ij} = \overset{1}{\gamma}_1 \overset{2}{\gamma}_1 \alpha_{11} + \overset{1}{\gamma}_2 \overset{2}{\gamma}_2 \alpha_{22} + \overset{1}{\gamma}_1 \overset{2}{\gamma}_2 \alpha_{12} + \overset{1}{\gamma}_2 \overset{2}{\gamma}_1 \alpha_{21} \\ &= \alpha_{11} + \alpha_{22} + \mathrm{i}(\alpha_{12} - \alpha_{21}).\end{aligned} \tag{9.125}$$

We apply the same transformation rules to the stress tensor components τ^{11}, and τ^{12}, using that the stress tensor is symmetric,

$$\tau^{11} = \sigma_{11} - \sigma_{22} - 2\mathrm{i}\sigma_{12}, \tag{9.126}$$

and

$$\tau^{12} = \sigma_{11} + \sigma_{22}. \tag{9.127}$$

We transform the strain tensor the same way:

$$S^{11} = e_{11} - e_{22} - 2\mathrm{i}e_{12}, \tag{9.128}$$

and

$$S^{12} = e_{11} + e_{22}. \tag{9.129}$$

We combine the first two equations of Hooke's Law, (9.115) and (9.116) into a single complex one

$$\boxed{\tau^{11} = 2GS^{11}}, \tag{9.130}$$

and transform (9.117) :

$$\boxed{\tau^{12} = \frac{2G}{1-2\nu} S^{12}}. \tag{9.131}$$

Equilibrium Conditions

The equilibrium conditions in terms of Cartesian coordinates are given by (5.87) on page 65, which we derived for the case that tensile stresses are positive. We consider the case of static equilibrium of a geotechnical material, such as rock, and take compression positive. This results in the following equation:

$$\partial_m \sigma_{mn} = \beta_n, \tag{9.132}$$

where β_n represents the body force. We transform the first one of these equations in terms of complex variables,

$$\tilde{\partial}_m \tau^{m1} = \frac{\partial \tau^{11}}{\partial \bar{z}} + \frac{\partial \tau^{21}}{\partial z} = \tilde{\beta}^1. \tag{9.133}$$

The second equilibrium equation, $n = 2$, is the complex conjugate of the first, and is superfluous. We represent the transformed body force as B_0,

$$\tilde{\beta}^1 = B_0 = \beta_x - \mathrm{i}\beta_y. \tag{9.134}$$

We use Hooke's law to express the stress tensor components in terms of the strains, and combine the result with the equilibrium equation,

$$2G \frac{\partial S^{11}}{\partial \bar{z}} + \frac{2G}{1-2\nu} \frac{\partial S^{12}}{\partial z} = B_0. \tag{9.135}$$

The strain tensor component S^{11} is minus the component $\tilde{\alpha}_{11}$ of the displacement gradient tensor,

$$S^{11} = -\tilde{\alpha}^{11} = -\tilde{\partial}^1 \tilde{u}^1; \tag{9.136}$$

the minus sign occurs because we take compression positive. The component $\tilde{u}^1$ equals w, and $\tilde{\partial}^1 = 2\partial/\partial z$ according to (8.57) on page 92,

$$S^{11} = -2\frac{\partial w}{\partial z}. \tag{9.137}$$

The strain tensor is minus the symmetrical part of the displacement gradient tensor,

$$S^{12} = -\tfrac{1}{2}(\tilde{\alpha}^{12} + \tilde{\alpha}^{21}) = -\tfrac{1}{2}(\tilde{\partial}^1 \tilde{u}^2 + \tilde{\partial}^2 \tilde{u}^1) = -\left(\frac{\partial \bar{w}}{\partial z} + \frac{\partial w}{\partial \bar{z}}\right). \tag{9.138}$$

We combine the latter two equations with (9.135),

$$4G\frac{\partial^2 w}{\partial z\partial \bar{z}} + \frac{2G}{1-2\nu}\left(\frac{\partial^2 \bar{w}}{\partial z^2} + \frac{\partial^2 w}{\partial z\partial \bar{z}}\right) = -B_0, \tag{9.139}$$

and combine terms,

$$\left(4G + \frac{2G}{1-2\nu}\right)\frac{\partial^2 w}{\partial z\partial \bar{z}} + \frac{2G}{1-2\nu}\frac{\partial^2 \bar{w}}{\partial z^2} = \frac{2G}{1-2\nu}(3-4\nu)\frac{\partial^2 w}{\partial z\partial \bar{z}} + \frac{2G}{1-2\nu}\frac{\partial^2 \bar{w}}{\partial z^2} = -B_0 \tag{9.140}$$

We multiply both sides by $(1-2\nu)/2G$,

$$\kappa\frac{\partial^2 w}{\partial z\partial \bar{z}} + \frac{\partial^2 \bar{w}}{\partial z^2} = -\frac{1-2\nu}{2G}B_0, \tag{9.141}$$

where

$$\kappa = 3-4\nu. \tag{9.142}$$

We apply the Helmholtz decomposition theorem, which states that any vector field can be written as the sum of the gradient of a scalar potential and the curl of a vector potential, which implies in two dimensions that we may write w as

$$w = -2\frac{\partial \Gamma}{\partial z}, \tag{9.143}$$

where Γ is some complex function. We substitute this expression for w in (9.141),

$$\kappa\frac{\partial^3 \Gamma}{\partial z^2\partial \bar{z}} + \frac{\partial^3 \bar{\Gamma}}{\partial z^2\partial \bar{z}} = \frac{1-2\nu}{4G}B_0, \tag{9.144}$$

differentiate both sides with respect to $\bar{z}$,

$$\kappa\frac{\partial^4 \Gamma}{\partial z^2\partial \bar{z}^2} + \frac{\partial^4 \bar{\Gamma}}{\partial z^2\partial \bar{z}^2} = \frac{1-2\nu}{4G}\frac{\partial B_0}{\partial \bar{z}}, \tag{9.145}$$

and combine terms:

$$\frac{\partial^4(\kappa\Gamma + \bar{\Gamma})}{\partial z^2\partial \bar{z}^2} = \frac{1-2\nu}{4G}\frac{\partial B_0}{\partial \bar{z}}. \tag{9.146}$$

We restrict our analysis to harmonic body forces, whose gradients are both irrotational and divergence-free:

$$\frac{\partial B_0}{\partial \bar{z}} = 0, \tag{9.147}$$

Examples of such body forces are gravitational forces, which are constant, and seepage forces caused by groundwater flow; (9.146) reduces to

$$\frac{\partial^4(\kappa\Gamma + \bar{\Gamma})}{\partial z^2\partial \bar{z}^2} = 0. \tag{9.148}$$

The operator $\partial^4/(\partial z^2\partial \bar{z}^2)$ is real; a complex function that is equal to a factor that is unequal to one, i.e., for $\nu \neq 1/2$, times its complex conjugate is zero,

$$\frac{\partial^4 \Gamma}{\partial z^2\partial \bar{z}^2} = 0. \tag{9.149}$$

9.5.2 General Solution

We integrate (9.149) first with respect to z and then with respect to $\bar{z}$ and multiply the result by -2 to obtain an expression for $\partial w/\partial \bar{z}$, which gives

$$-2\frac{\partial^2 \Gamma}{\partial z \partial \bar{z}} = \frac{\partial w}{\partial \bar{z}} = f_0(z) + g_0(\bar{z}) \tag{9.150}$$

where $f_0(z)$ and $g_0(\bar{z})$ are arbitrary complex functions. We integrate (9.150) with respect to $\bar{z}$,

$$w = \bar{z} f_0(z) + g_1(\bar{z}) + h_0(z), \tag{9.151}$$

where $h_0(z)$ is an arbitrary function of z, and

$$g_1(\bar{z}) = \int g_0(\bar{z}) d\bar{z}. \tag{9.152}$$

We lost information by differentiating (9.144); we regain this information by substituting (9.151) for w into (9.141), which requires expressions for $\partial^2 w/(\partial z \partial \bar{z})$ and $\partial^2 \bar{w}/\partial z^2$, which we determine in what follows. We obtain from (9.151)

$$\frac{\partial w}{\partial z} = \bar{z} f_0'(z) + h_0'(z), \quad \frac{\partial^2 w}{\partial z \partial \bar{z}} = f_0'(z), \tag{9.153}$$

where the prime stands for differentiation of the function with respect to the single variable it operates on. We take the complex conjugate of (9.151),

$$\bar{w} = z \bar{f}_0(\bar{z}) + \bar{g}_1(z) + \bar{h}_0(\bar{z}). \tag{9.154}$$

We placed a bar over letters f, g, and h; we define this operation for an arbitrary function $f(z)$ as

$$\bar{f}(z) = \overline{f(\bar{z})}, \tag{9.155}$$

so that $\bar{f}$ is the conjugate operation of f; if f is a real operator, then $\bar{f} = f$. If we take the complex conjugate of both sides of (9.155), we find that

$$\overline{\bar{f}(z)} = f(\bar{z}). \tag{9.156}$$

We differentiate (9.154) with respect to z twice,

$$\frac{\partial \bar{w}}{\partial z} = \bar{f}_0(\bar{z}) + \bar{g}_0(z), \quad \frac{\partial^2 \bar{w}}{\partial z^2} = \bar{g}_0'(z), \tag{9.157}$$

where $g_1' = g_0$. We use (9.157) and (9.153) in (9.141),

$$\kappa f_0'(z) + \bar{g}_0'(z) = -\frac{1-2\nu}{2G} B_0. \tag{9.158}$$

Expression (9.151) for w contains the function $g_1(\bar{z})$, which we eliminate using (9.158). We integrate the latter equation twice to obtain an expression for $\bar{g}_1(z)$. The first integration gives

$$\kappa f_0(z) + \bar{g}_0(z) = -\frac{1-2\nu}{2G} B_1, \tag{9.159}$$

where

$$B_n = \int B_{n-1} dz, \quad n = 1, 2, \cdots. \tag{9.160}$$

We integrate (9.159) with respect to z,

$$\kappa f_1(z) + \bar{g}_1(z) = -\frac{1-2\nu}{2G} B_2. \tag{9.161}$$

We solve this for g_1 and take the complex conjugate of the result,

$$\overline{\bar{g}_1(z)} = g_1(\bar{z}) = -\kappa \bar{f}_1(\bar{z}) - \frac{1-2\nu}{2G} \bar{B}_2, \tag{9.162}$$

and substitute this for $g_1(\bar{z})$ in (9.151) to express w in terms of two unknown functions, to be determined from the boundary conditions:

$$w = \bar{z} f_0(z) - \kappa \bar{f}_1(\bar{z}) + h_0(z) - \frac{1-2\nu}{2G} \bar{B}_2, \tag{9.163}$$

and

$$\bar{w} = z \bar{f}_0(\bar{z}) - \kappa f_1(z) + \bar{h}_0(\bar{z}) - \frac{1-2\nu}{2G} B_2. \tag{9.164}$$

The governing equation for linear elasticity is biharmonic and requires two boundary values along each section of the boundary.

9.5.3 Stresses and Strains

We obtain expressions for the strains S^{11} and S^{12} by the use of (9.137), (9.138) on page 137, and (9.162)

$$S^{11} = -2\frac{\partial w}{\partial z} = -2\left[\bar{z} f_0'(z) + h_0'(z)\right], \tag{9.165}$$

$$S^{12} = -\left(\frac{\partial \bar{w}}{\partial z} + \frac{\partial w}{\partial \bar{z}}\right) = -\left(\bar{f}_0(\bar{z}) + f_0(z) - \kappa f_0(z) - \kappa \bar{f}_0(\bar{z}) - \frac{1-2\nu}{2G}(B_1 + \bar{B}_1)\right) \tag{9.166}$$

or

$$S^{12} = (\kappa - 1)[f_0(z) + \bar{f}_0(\bar{z})] + \frac{1-2\nu}{2G}\left(\bar{B}_1 + B_1\right). \tag{9.167}$$

We use these expressions for the strains in Hooke's law, given by (9.130) on page 136,

$$\tau^{11} = -4G\left[\bar{z} f_0'(z) + h_0'(z)\right], \tag{9.168}$$

and (9.131), with $\kappa = 3 - 4\nu$, so that $\kappa - 1 = 2(1-2\nu)$,

$$\tau^{12} = \frac{2G}{1-2\nu} S^{12} = \frac{2G}{1-2\nu}\{2(1-2\nu)[f_0(z) + \bar{f}_0(\bar{z})]\} + \left(\bar{B}_1 + B_1\right), \tag{9.169}$$

or

$$\tau^{12} = 4G[f_0(z) + \bar{f}_0(\bar{z})] + \left(\bar{B}_1 + B_1\right). \tag{9.170}$$

We introduce new functions $\Phi(z)$ and $\Psi(z)$,

$$\Phi'(z) = -4Gf_0(z), \tag{9.171}$$

and

$$\Psi(z) = 4Gh_0(z), \tag{9.172}$$

and express w, (9.163), τ^{11}, (9.168), and τ^{12}, (9.169), in terms of these functions:

$$w = \frac{1}{4G}\{-\bar{z}\Phi'(z) + \kappa\bar{\Phi}(\bar{z}) + \Psi(z) - 2(1-2\nu)\bar{B}_2\}, \tag{9.173}$$

and

$$\tau^{11} = \bar{z}\Phi''(z) - \Psi'(z), \tag{9.174}$$

and

$$\tau^{12} = -\Phi'(z) - \bar{\Phi}'(\bar{z}) + (\bar{B}_1 + B_1). \tag{9.175}$$

These equations are equivalent to the Muskhelishvili-Kolosov functions, see Muskhelishvili [1953].

9.5.4 Tractions and Displacements Acting on Arbitrary Planes

Boundary conditions are usually expressed in terms of displacements and/or components of the traction on a given plane. The angle α is between the plane and the x_1-axis, such that the outward normal to the plane points at ninety degrees to the left with respect to a vector in the direction α, see Figure 9.19. If $\underset{\alpha}{w}$ represents the complex

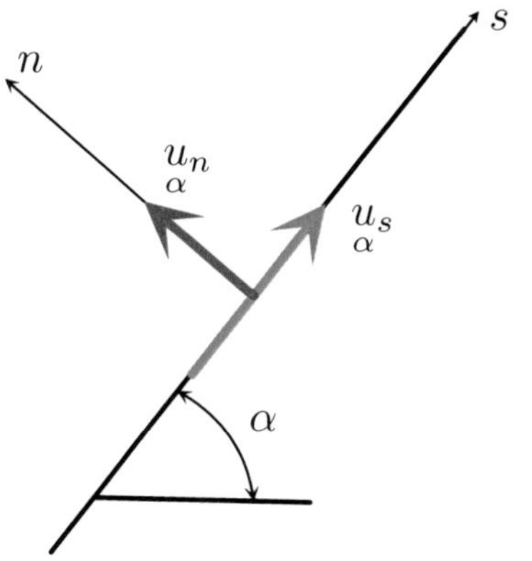

Figure 9.19: Displacements on a plane of orientation α.

displacement in local coordinates s, n, with s in the direction α and n in the direction of the normal to the plane, then[2]

$$\underset{\alpha}{w} = \underset{\alpha}{u_s} - \mathrm{i}\underset{\alpha}{u_n} = w\mathrm{e}^{\mathrm{i}\alpha}, \tag{9.176}$$

[2]Multiplication of a complex number by $\mathrm{e}^{\mathrm{i}\alpha}$ is equivalent to rotating the line from the origin to the point representing the complex number over an angle α in counterclockwise direction.

where $\underset{\alpha}{u_s}$ and $\underset{\alpha}{u_n}$ represent the tangential and normal displacements, respectively. We obtain from (9.176) with (9.173)

$$\boxed{w = \frac{e^{i\alpha}}{4G}\{-\bar{z}\Phi'(z) + \kappa\bar{\Phi}(\bar{z}) + \Psi(z) - 2(1-2\nu)\bar{B}_2\}}. \tag{9.177}$$

We obtain expressions for the tractions $\underset{\alpha}{t_s}$ and $\underset{\alpha}{t_n}$ in a similar manner. The Cartesian components of the traction in terms of stresses on a plane with normal n_q are

$$t_p = -\sigma_{pq}n_q, \tag{9.178}$$

where the $-$ sign occurs because we take compression positive. We transform this into

$$T^p = -\tau^{pq}N_q, \tag{9.179}$$

where N_q is the contravariant unit normal,

$$N_1 = \tfrac{1}{2}(n_x + in_y) = \tfrac{1}{2}e^{i(\alpha+\pi/2)} \qquad N_2 = \bar{N}_1. \tag{9.180}$$

We write $\underset{\alpha}{T^1} = \underset{\alpha}{t_s} - i\underset{\alpha}{t_n}$ in terms of T^1:

$$\underset{\alpha}{T^1} = \underset{\alpha}{t_s} - i\underset{\alpha}{t_n} = T^1 e^{i\alpha}, \tag{9.181}$$

where (s,n) are the local coordinates. We combine (9.179), (9.180), and (9.181),

$$\underset{\alpha}{T^1} = \underset{\alpha}{t_s} - i\underset{\alpha}{t_n} = -e^{i\alpha}[\tau^{11}N_1 + \tau^{12}N_2] = -\tfrac{1}{2}e^{i\alpha}\left[\tau^{11}e^{i(\alpha+\pi/2)} + \tau^{12}e^{-i(\alpha+\pi/2)}\right], \tag{9.182}$$

or

$$\boxed{\underset{\alpha}{T^1} = -\tfrac{1}{2}i\left[\tau^{11}e^{2i\alpha} - \tau^{12}\right]}. \tag{9.183}$$

We apply these equations in what follows to obtain solutions to several problems in linear elasticity.

9.5.5 Stresses Acting On a Half Plane

We consider several cases of stresses that act on a half plane, at first ignoring body forces. The boundary of the domain is a straight line, and we can choose the function Ψ such that the terms that contain the factor $\bar{z}$ vanish along the boundary. For the special case that the straight boundary is defined by $y = 0$, the choice for Ψ is:

$$\Psi = z\Phi'(z) + \psi(z), \tag{9.184}$$

where ψ is a newly introduced function. We differentiate Ψ:

$$\Psi'(z) = z\Phi''(z) + \Phi'(z) + \psi'(z), \tag{9.185}$$

and use this in expression (9.177) for w, with $\alpha = 0$ and $\bar{B}_2 = 0$,

$$w = \frac{1}{4G}\left\{(z-\bar{z})\Phi'(z) + \kappa\bar{\Phi}(\bar{z}) + \psi(z)\right\}, \tag{9.186}$$

and in expression (9.174) for τ^{11},

$$\tau^{11} = -(z-\bar{z})\Phi''(z) - \Phi'(z) - \psi'(z); \tag{9.187}$$

the expression for τ^{12} remains the same.

Normal load along $x \leq 0$

The half plane $y \leq 0$ is loaded along $x < 0, y = 0$ as illustrated in Figure 9.20. We do not include the effect of gravity. The boundary conditions are :

$$t_x = 0, \quad t_y = -p \quad x \leq 0,\ y = 0, \tag{9.188}$$

and

$$t_x = 0, \quad t_y = 0 \quad x > 0,\ y = 0. \tag{9.189}$$

We apply (9.182) with $\alpha = 0$,

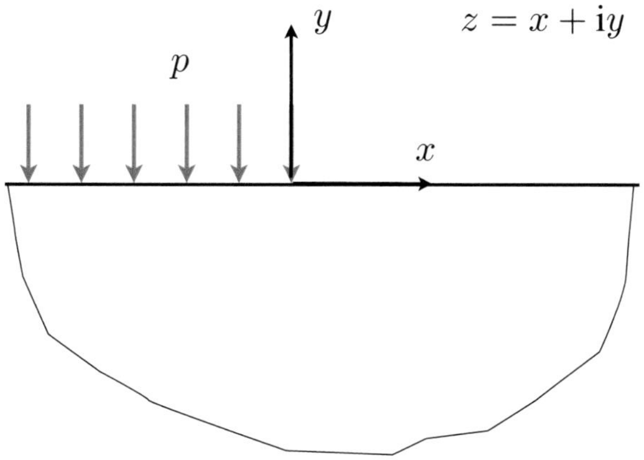

Figure 9.20: A half space loaded by a normal stress p along $x \leq 0$

$$\begin{aligned} T^1 &= t_x - \mathrm{i}t_y = -\tfrac{1}{2}\mathrm{i}\left[\tau^{11} - \tau^{12}\right] = \mathrm{i}p \qquad x \leq 0,\ y = 0 \\ T^1 &= t_x - \mathrm{i}t_y = -\tfrac{1}{2}\mathrm{i}\left[\tau^{11} - \tau^{12}\right] = 0 \qquad x > 0,\ y = 0, \end{aligned} \tag{9.190}$$

and use the expressions (9.187) and (9.175) for the stresses,

$$\begin{aligned} T^1 &= t_x - \mathrm{i}t_y = \tfrac{1}{2}\mathrm{i}\left[(z-\bar{z})\Phi''(z) + \psi'(z) - \bar{\Phi}'(\bar{z})\right] = \mathrm{i}p \qquad x \leq 0,\ y = 0 \\ T^1 &= t_x - \mathrm{i}t_y = \tfrac{1}{2}\mathrm{i}\left[(z-\bar{z})\Phi''(z) + \psi'(z) - \bar{\Phi}'(\bar{z})\right] = 0 \qquad x > 0,\ y = 0. \end{aligned} \tag{9.191}$$

Since $z = \bar{z}$ along $y = 0$, the terms with the factor $(z - \bar{z})$ disappear along the boundary. The traction must be either zero or purely imaginary along the boundary, depending on whether x is greater than or less than zero; the term inside the brackets in (9.191) must be purely real, i.e.,

$$\psi'(z) = -\Phi'(z). \tag{9.192}$$

The real part of this function must be equal to $-p$ for $x < 0, y = 0^-$, and 0 for $x \geq 0, y = 0^-$.[3] The function $\psi'(z)$ that satisfies this condition is:

$$\Phi' = -\psi' = \frac{p}{\pi \mathrm{i}} \ln(z). \tag{9.193}$$

We use this in (9.191), for $x \leq 0, y = 0$, i.e., for $z = |z| e^{-\mathrm{i}\pi}$, and[4] $z = \bar{z}$

$$T^1 = t_x - \mathrm{i}t_y = \tfrac{1}{2}\mathrm{i}\left[-\frac{p}{\pi \mathrm{i}}[\ln|z| - \mathrm{i}\pi] - \frac{p}{\pi(-\mathrm{i})}[\ln|z| + \mathrm{i}\pi] \right] = \mathrm{i}p \rightarrow t_y = -p. \tag{9.194}$$

We set $x > 0, y = 0^-$ and $z = |z|$:

$$T^1 = t_x - \mathrm{i}t_y = \tfrac{1}{2}\mathrm{i}\left[-\frac{p}{\pi \mathrm{i}} \ln|z| - \frac{p}{\pi(-\mathrm{i})} \ln|z| \right] = 0 \rightarrow t_y = -p. \tag{9.195}$$

We integrate (9.193) to find the function Φ,

$$\Phi = \frac{p}{\pi \mathrm{i}}[z \ln(z) - z] + C, \tag{9.196}$$

where C is a constant.

Displacements

We obtain the expression for the complex displacements w from (9.177) by setting both the angle α and the body force B equalto zero:

$$w = u_x - \mathrm{i}u_y = \frac{1}{4G}\left\{ -\bar{z}\Phi'(z) + \kappa\bar{\Phi}(\bar{z}) + \Psi(z) \right\}. \tag{9.197}$$

Recall that we replaced the function Ψ by $z\Phi'(z) + \psi(z)$, see (9.184), and that we found that $\psi'(z) = -\Phi'(z)$,

$$w = \frac{1}{4G}\left\{ (z - \bar{z})\Phi'(z) - \Phi(z) + \kappa\bar{\Phi}(\bar{z}) \right\}. \tag{9.198}$$

We use (9.193) and (9.196) to obtain an expression for w as a function of position:

$$w = \frac{1}{4G}\left\{ \frac{p}{\pi \mathrm{i}}(z - \bar{z}) \ln(z) - \frac{p}{\pi \mathrm{i}}[z \ln(z) - z] - C - \kappa \frac{p}{\pi \mathrm{i}}[\bar{z} \ln(\bar{z}) - \bar{z}] + \kappa \bar{C} \right\}, \tag{9.199}$$

or

$$w = -\frac{p}{4G\pi \mathrm{i}}\{ -(z - \bar{z}) \ln(z) + [z \ln(z) - z] + \kappa[\bar{z} \ln(\bar{z}) - \bar{z}] \} + w_\infty, \tag{9.200}$$

[3] The $-$ in 0^- indicates that the domain of interest is the lower half plane, where the argument of z is $-\pi$ rather than π.

[4] The argument of z is $-\mathrm{i}\pi$ because we consider the lower half plane, so that the argument of z falls between 0 and $-\pi$.

where w_∞ is a new constant with the dimension of length.

$$w_\infty = \frac{\kappa \bar{C} - C}{4G}. \tag{9.201}$$

The displacements increase toward infinity, and are not bounded; the increase is logarithmic, and rather slow. We compute the constant w_∞ by setting the displacement components equal to zero at some point $x = L$.

This solution is not useful by itself as the load extends to infinity. Combination of this solution with another one gives us the solution for a strip load, which is useful.

Stress Tensor Components
The stress tensor component τ^{11} is given by (9.174),

$$\tau^{11} = \bar{z}\Phi''(z) - \Psi'(z). \tag{9.202}$$

We use (9.185), $\Psi' = z\Phi''(z) + \Phi'(z) + \psi'(z)$ with $\Phi'(z) = -\psi'(z) = p/(\pi \mathrm{i}) \ln(z)$,

$$\tau^{11} = -(z - \bar{z})\Phi''(z) = -\frac{p}{\pi \mathrm{i}} \left\{ \frac{z - \bar{z}}{z} \right\}. \tag{9.203}$$

We obtain an expression for the stress tensor component τ^{12} from (9.175), setting the body forces equal to zero,

$$\tau^{12} = -\Phi'(z) - \bar{\Phi}'(\bar{z}) = -\frac{p}{\pi \mathrm{i}} \left[\ln(z) - \ln(\bar{z}) \right]. \tag{9.204}$$

We express the transformed tensor components in terms of the components in Cartesian space using (9.126) and (9.127) on page 136,

$$\begin{aligned} \tau^{11} &= \sigma_{11} - \sigma_{22} - 2\mathrm{i}\sigma_{12} \\ \tau^{12} &= \sigma_{11} + \sigma_{22}. \end{aligned} \tag{9.205}$$

We obtain expressions for the individual stress components from:

$$\sigma_{11} - \mathrm{i}\sigma_{12} = \tfrac{1}{2}\left[\tau^{11} + \tau^{12}\right] = \frac{p}{2\pi \mathrm{i}} \left\{ -\frac{z - \bar{z}}{z} - \ln(z) + \ln(\bar{z}) \right\} \tag{9.206}$$

$$\sigma_{22} + \mathrm{i}\sigma_{12} = \tfrac{1}{2}\left[-\tau^{11} + \tau^{12}\right] = \frac{p}{2\pi \mathrm{i}} \left\{ \frac{z - \bar{z}}{z} - \ln z + \ln \bar{z} \right\}. \tag{9.207}$$

Problem 9.2

Consider the case of a semi infinite load on a half space.

Questions:

1. Set the displacement components at some point $z = L$, where L is real, to zero and compute the constant w_∞ from that condition.
2. Verify that all boundary conditions are met.
3. Produce a plot of the two displacement components along:

(a) The boundary $y = 0$,

(b) A line at a distance d of your choice below the surface.

Problem 9.3

Produce a plot of the stress tensor components $\sigma_{22} = \sigma_{yy}$ and $\sigma_{12} = \sigma_{xy}$ along a line at a distance d below the surface.

1. The boundary $y = 0$,
2. A line at a distance d of your choice below the surface.

9.5.6 A Half-Space Loaded Along a Section $-a \leq x \leq a$ on the Surface $y = 0$

We obtain the solution for a half space loaded by a vertical force $t_y = -p$ along the section $-a \leq x \leq a$ by combining two of the solutions presented for the case of a semi-infinite load, see Figure 9.21. We modify the solutions by moving the origin to a point $z = a$, which gives, with (9.200)

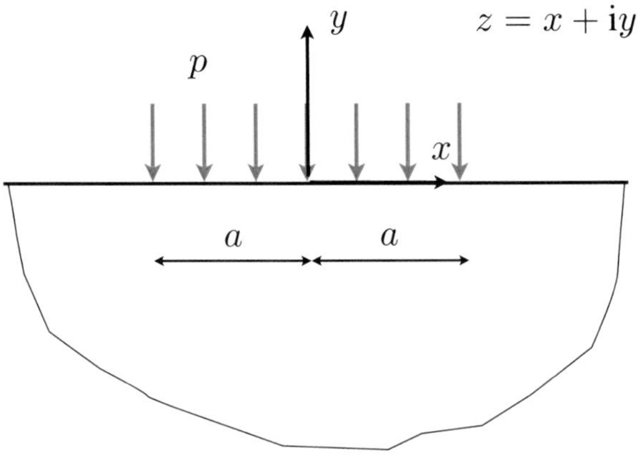

Figure 9.21: A half-space with a strip load

$$w = -\frac{p}{4G\pi \mathrm{i}}\Big\{ -(z-\bar{z})\ln(z-a) + [(z-a)\ln(z-a)-(z-a)] \\ + \kappa[(\bar{z}-a)\ln(\bar{z}-a)-(\bar{z}-a)]\Big\} + w_\infty \tag{9.208}$$

This solution corresponds to the case that the half-plane is loaded by a vertical load $t_y = -p$ from $z = -\infty$ to $z = a$. We add a second solution, corresponding to a negative load from $z = -\infty$ to $z = -a$. This second solution creates a vertical traction $t_y = p$ from $z = -a$ to $z = -\infty$, canceling the vertical traction from $z = -a$ to $z = -\infty$, leaving

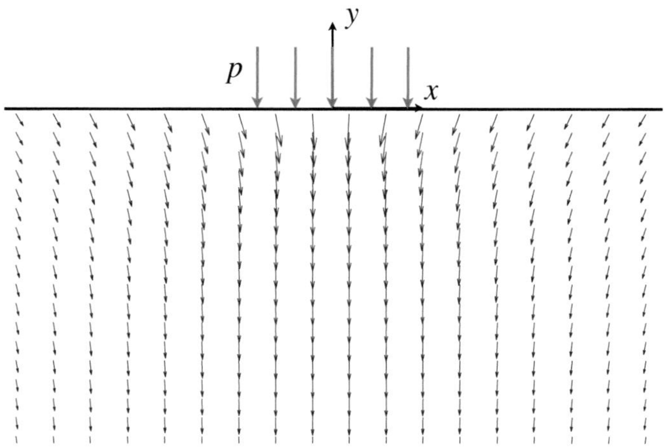

Figure 9.22: Displacements below a strip load over a width $2a$ on a half space

only the section $-a \leq z \leq a$ with a vertical load of $t_y = -p$. The result is:

$$\begin{aligned} w = &-\frac{p}{4G}\frac{1}{\pi \mathrm{i}}\Big\{ -(z-\bar{z})\ln\frac{z-a}{z+a} + [(z-a)\ln(z-a)-(z-a)] \\ &-[(z+a)\ln(z+a)-(z+a)] + \kappa\Big[(\bar{z}-a)\ln(\bar{z}-a)-(\bar{z}-a) \\ &-[(\bar{z}+a)\ln(\bar{z}+a)-(\bar{z}+a)]\Big] + C \end{aligned} \tag{9.209}$$

where C is a constant. We rearrange (9.209)

$$\begin{aligned} w = &-\frac{p}{4G}\frac{1}{\pi \mathrm{i}}\Big\{ -(z-\bar{z})\ln\frac{z-a}{z+a} + (z-a)\ln(z-a)-(z+a)\ln(z+a)] \\ &+\kappa[(\bar{z}-a)\ln(\bar{z}-a)-(\bar{z}+a)\ln(\bar{z}+a)]\Big\} + w_0 \end{aligned} \tag{9.210}$$

where w_0 is a new constant, which combines a constant that depends on a with C. We obtain expressions for the stresses in a similar way by superposition.

Displacements are shown in Figure 9.22 for the case that $a = 2$, $\nu = .3$, and $p/G = 0.1$. The three components of the stress tensor computed along a line $x - ia$ are shown in Figure 9.23. A plot of the major principal stress is shown in Figure 9.24. The lines are inclined as the principal directions, and the length of each line is proportional to the magnitude of the stress.

Problem 9.4

This problem concerns a strip load on a half-space.

Questions:

1. Determine an expression for the displacements along the boundary at large distances from the origin.

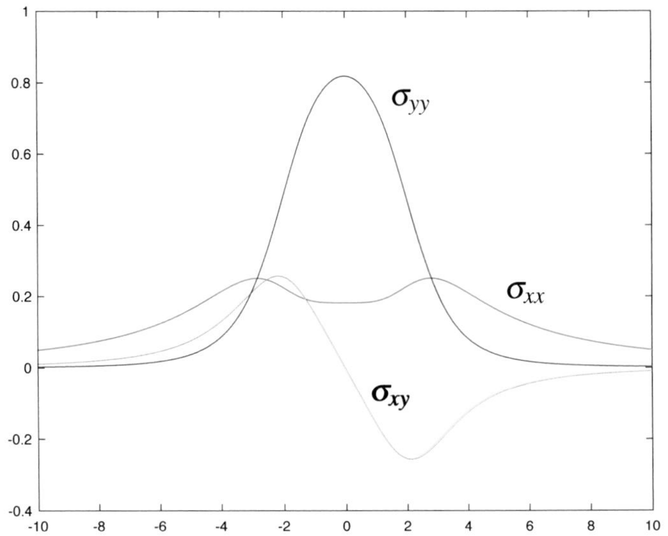

Figure 9.23: The three stress tensor components along the plane $z = x - ia$.

2. Present expressions for the three components of the stress tensor in Cartesian space.
3. Verify that all boundary conditions are met.
4. Produce a plot of the two displacement components along:
   (a) The boundary $y = 0$,
   (b) A line at a distance d of your choice below the surface.

Problem 9.5

This problem concerns a strip load on the boundary of a half space. Produce a plot of the stress tensor components $\sigma_{22} = \sigma_{yy}$ and $\sigma_{12} = \sigma_{xy}$ along :

1. The boundary $y = 0$,
2. A line at a distance d of your choice below the surface.

Problem 9.6

This problem concerns a strip load on the boundary of a half space.

Questions:

1. Present an expression for the complex combination of divergence and curl, ω.
2. Produce a contour plot of the real and imaginary parts of the function, and comment on the results.

9.5.7 Flamant's Problem

Flamant's problem concerns a concentrated load at a point on a half space, a point load. We obtain expressions for both the displacement field and the stresses by taking

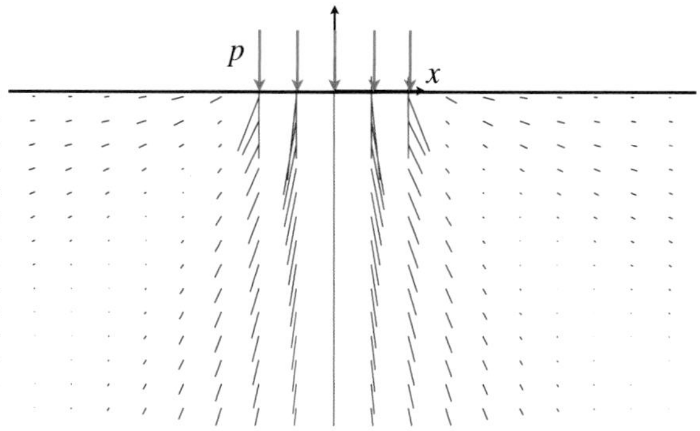

Figure 9.24: Plot of the major stresses below a strip load. The lengths of the lines are proportional to the magnitude of the stress

the limit of the solution for a strip load for the width a approaching zero while the traction p approaches infinity,

$$\lim_{\substack{p\to\infty\\ a\to 0}} = 2pa = F, \tag{9.211}$$

where F is the force. We apply this approach to the expression for w, (9.210), and rewrite it in a form suitable for taking the limit, introducing a new constant C,

$$\begin{aligned} w = -\lim_{\substack{p\to\infty\\ a\to 0}} \frac{p}{4G}\frac{1}{\pi \mathrm{i}}\Bigg\{ &-(z-\bar{z})\ln\frac{z-a}{z+a} + z\ln\frac{z-a}{z+a} - a\ln(z^2-a^2) \\ &+\kappa\left[\bar{z}\ln\frac{\bar{z}-a}{\bar{z}+a} - a\ln(\bar{z}^2-a^2)\right]\Bigg\} + C, \end{aligned} \tag{9.212}$$

or

$$\begin{aligned} w = -\lim_{\substack{p\to\infty\\ a\to 0}} \frac{p}{4G}\frac{1}{\pi \mathrm{i}}\Bigg\{ &-(z-\bar{z})\ln\frac{1-a/z}{1+a/z} + z\ln\frac{1-a/z}{1+a/z} - 2a\ln z - a\ln(1-a^2/z^2) \\ &+\kappa\left[\bar{z}\ln\frac{1-a/\bar{z}}{1+a/\bar{z}} - 2a\ln\bar{z} - a\ln(1-a^2/\bar{z}^2)\right]\Bigg\} + C. \end{aligned} \tag{9.213}$$

We expand $\ln(1+\varepsilon)$ about $\varepsilon = 0$, taking only the first term into account and take the limit

$$w = -\frac{F}{4G}\frac{1}{\pi \mathrm{i}}\left\{\frac{z-\bar{z}}{z} - 1 - \ln z + \kappa[-1-\ln\bar{z}]\right\} + C. \tag{9.214}$$

The component u_x must be symmetrical with respect to the origin; we may add a constant displacement, since this does not affect the strains. We modify the constant such that the real part of this constant, which contributes to the displacement u_x, is symmetrical with respect to the origin. The following expression for w satisfies this condition:

$$w = -\frac{F}{4G}\frac{1}{\pi \mathrm{i}}\left[\frac{z-\bar{z}}{z} - \ln z - \kappa\ln\bar{z} - \frac{\pi}{2}\mathrm{i}(1-\kappa)\right] - \mathrm{i}u_{y0}, \tag{9.215}$$

where the constant u_{y0} can be chosen to set the vertical displacement component equal to some given value at some chosen point. A plot of the displacements is shown in Figure 9.25 for the case that $F/G = 1$.

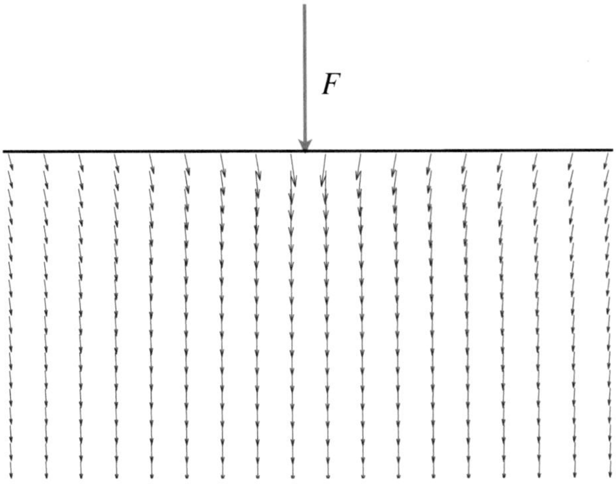

Figure 9.25: Displacements for Flamant's problem

Stresses

We obtain the components of the strain tensor from equations (9.137) and (9.138) on page 137 upon differentiation of the expression for w:

$$S^{11} = -2\frac{\partial w}{\partial z} = \frac{F}{2G}\frac{1}{\pi \mathrm{i}}\left[\frac{\bar{z}}{z^2} - \frac{1}{z}\right], \tag{9.216}$$

and

$$S^{12} = -\left[\frac{\partial w}{\partial \bar{z}} + \frac{\partial \bar{w}}{\partial z}\right] = \frac{F}{4G}\frac{1}{\pi \mathrm{i}}\left[-\frac{1}{z} - \kappa\frac{1}{\bar{z}} + \frac{1}{\bar{z}} + \kappa\frac{1}{z}\right] = \frac{F}{4G}\frac{1}{\pi \mathrm{i}}(\kappa - 1)\left[\frac{1}{z} - \frac{1}{\bar{z}}\right]. \tag{9.217}$$

We apply Hooke's law, equations (9.130) and (9.131) on page 137, and the identity $\kappa - 1 = 2 - 4\nu$ to obtain expressions for the stress tensor components:

$$\tau^{11} = \frac{F}{\pi \mathrm{i}}\left[\frac{\bar{z}}{z^2} - \frac{1}{z}\right] = -\frac{F}{\pi \mathrm{i}}\frac{z - \bar{z}}{z^2}, \tag{9.218}$$

and

$$\tau^{12} = \frac{F}{\pi \mathrm{i}}\left[\frac{1}{z} - \frac{1}{\bar{z}}\right]. \tag{9.219}$$

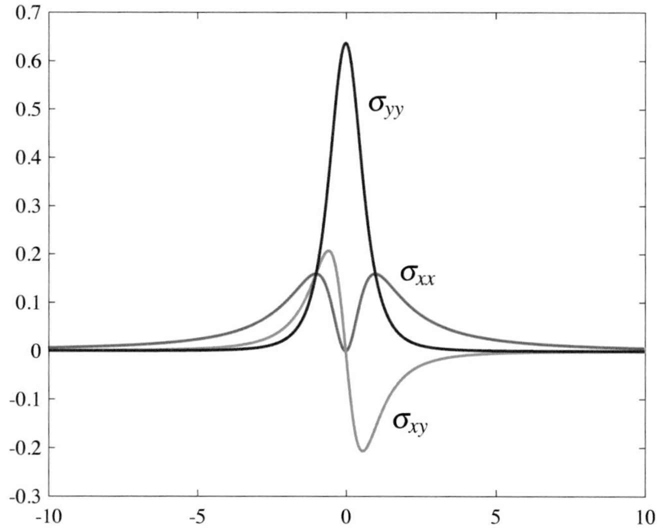

Figure 9.26: The three stress tensor components for Flamant's problem

We find expressions for the components of the stress tensor from (9.126) and (9.127) on page 136:

$$\sigma_{11} - \mathrm{i}\sigma_{12} = \tfrac{1}{2}(\tau^{11} + \tau^{12}) = \frac{F}{2\pi\mathrm{i}}\left[-\frac{z-\bar{z}}{z^2} + \frac{1}{z} - \frac{1}{\bar{z}}\right] \tag{9.220}$$

$$\sigma_{22} + \mathrm{i}\sigma_{12} = \tfrac{1}{2}(-\tau^{11} + \tau^{12}) = \frac{F}{2\pi\mathrm{i}}\left[\frac{z-\bar{z}}{z^2} + \frac{1}{z} - \frac{1}{\bar{z}}\right]. \tag{9.221}$$

We write this in a form suitable for separating the real and imaginary parts:

$$\sigma_{11} - \mathrm{i}\sigma_{12} = \frac{F}{2\pi\mathrm{i}}\left[\frac{-(z-\bar{z})\bar{z}^2}{(z\bar{z})^2} + \frac{\bar{z}-z}{z\bar{z}}\right] = -\frac{F}{2\pi\mathrm{i}}\frac{z-\bar{z}}{(z\bar{z})^2}\left[\bar{z}^2 + z\bar{z}\right] = -\frac{F}{2\pi\mathrm{i}}\frac{z-\bar{z}}{(z\bar{z})^2}\bar{z}(z+\bar{z}) \tag{9.222}$$

and

$$\sigma_{22} + \mathrm{i}\sigma_{12} = \frac{F}{2\pi\mathrm{i}}\left[\frac{(z-\bar{z})\bar{z}^2}{(z\bar{z})^2} + \frac{\bar{z}-z}{z\bar{z}}\right] = \frac{F}{2\pi\mathrm{i}}\frac{z-\bar{z}}{(z\bar{z})^2}\left[\bar{z}^2 - z\bar{z}\right] = -\frac{F}{2\pi\mathrm{i}}\frac{z-\bar{z}}{(z\bar{z})^2}\bar{z}(z-\bar{z}) \tag{9.223}$$

or

$$\sigma_{11} - \mathrm{i}\sigma_{12} = -\frac{F}{2\pi\mathrm{i}}\frac{2\mathrm{i}y}{(z\bar{z})^2}(x-\mathrm{i}y)2x = -\frac{2F}{\pi}\frac{x^2y}{(z\bar{z})^2} + \mathrm{i}\frac{2F}{\pi}\frac{xy^2}{(z\bar{z})^2}, \tag{9.224}$$

and

$$\sigma_{22} + \mathrm{i}\sigma_{12} = -\frac{F}{2\pi\mathrm{i}}\frac{2\mathrm{i}y}{(z\bar{z})^2}(x-\mathrm{i}y)2\mathrm{i}y = -\frac{2F}{\pi}\frac{y^3}{(z\bar{z})^2} - \mathrm{i}\frac{2F}{\pi}\frac{xy^2}{(z\bar{z})^2}, \tag{9.225}$$

We separate real and imaginary parts:

$$\sigma_{11} = -\frac{2F}{\pi}\frac{x^2 y}{(z\bar{z})^2} \tag{9.226}$$

$$\sigma_{22} = -\frac{2F}{\pi}\frac{y^3}{(z\bar{z})^2} \tag{9.227}$$

$$\sigma_{12} = -\frac{2F}{\pi}\frac{xy^2}{(z\bar{z})^2}. \tag{9.228}$$

The components of the stress tensor along the plane $z = x - 2\mathrm{i}$ are shown in Figure 9.26.

9.5.8 An Analytic Element for Gravity

The body force due to gravity is B_0,

$$B_0 = \beta_x - i\beta_y = i\rho g \tag{9.229}$$

where ρ is the density and g the acceleration of gravity. We obtain the functions B_1 and B_2 upon integration with respect to z

$$B_1 = \mathrm{i}\rho g z \qquad B_2 = \tfrac{1}{2} i\rho g z^2. \tag{9.230}$$

9.5.9 A Pressurized Crack

Applying pressure inside a crack causes a discontinuity in displacement. We limit our analysis to the case that the crack falls along the real axis, and extends from $z = -L/2$ to $z = L/2$. The boundary conditions are along the real axis; we use the functions introduced for the case of a half plane. There is no shear stress inside the crack; the tractions along the two sides of the crack are:

$$\underset{\beta}{T}^1 = \underset{\beta}{t}_s - \mathrm{i}\underset{\beta}{t}_n = \mathrm{i}p, \tag{9.231}$$

where p is the pressure; the normal traction is negative as it acts against the direction of the outward normal. The pressure is the same on both sides of the crack; the components of the stress tensor are continuous across the crack. Since the crack opens as a result of the applied pressure, the displacements are discontinuous.

There are two boundary conditions along the crack: both the shear and normal components of the traction are given. The boundary condition at infinity is that the displacements vanish,

$$w \to 0 \qquad z \to \infty. \tag{9.232}$$

The exact solution for a pressurized crack is due to Griffith [1921].

The expression for w is given by (9.197) on page 144:

$$w = u_x - \mathrm{i}u_y = \frac{1}{4G}\left\{-\bar{z}\Phi'(z) + \kappa\bar{\Phi}(\bar{z}) + \Psi(z)\right\}, \tag{9.233}$$

and the expressions for τ^{11} and τ^{12} are given by (9.202) and (9.204):

$$\begin{aligned}\tau^{11} &= \bar{z}\Phi''(z) - \Psi'(z),\\ \tau^{12} &= -\Phi'(z) - \bar{\Phi}'(\bar{z}).\end{aligned} \tag{9.234}$$

We choose the following expression for Ψ,

$$\Psi = z\Phi' - \Phi, \tag{9.235}$$

so that

$$\Psi' = z\Phi'' + \Phi' - \Phi' = z\Phi'' \tag{9.236}$$

The crack extends from $z = -L/2$ to $z = L/2$, and we define the dimensionless variable Z as

$$Z = \frac{2}{L}z, \tag{9.237}$$

where L is the length of the crack. The expression for w becomes

$$w = \frac{1}{4G}\left\{\frac{L}{2}(Z-\bar{Z})\Phi'(z) + \kappa\bar{\Phi}(\bar{z}) - \Phi(z)\right\}, \tag{9.238}$$

and we obtain for τ^{11} and τ^{12},

$$\begin{aligned}\tau^{11} &= -\frac{L}{2}(Z-\bar{Z})\Phi''(z)\\ \tau^{12} &= -\Phi'(z) - \bar{\Phi}'(\bar{z})\end{aligned} \tag{9.239}$$

Exact Solution

We demonstrate that the following expression for Φ leads to a solution that meets the boundary conditions along the crack:

$$\Phi(Z) = \frac{pL}{2}\left[\sqrt{(Z+1)(Z-1)} - Z\right], \tag{9.240}$$

and

$$\bar{\Phi}(\bar{Z}) = \frac{pL}{2}\left[\sqrt{(\bar{Z}+1)(\bar{Z}-1)} - \bar{Z}\right], \tag{9.241}$$

We differentiate Φ with respect to z,

$$\Phi'(z) = \frac{pL}{2}\frac{2}{L}\left[\frac{Z}{\sqrt{(Z+1)(Z-1)}} - 1\right] = p\left[\frac{Z}{\sqrt{(Z+1)(Z-1)}} - 1\right], \tag{9.242}$$

and use this in (9.238) for w,

$$\begin{aligned}w = \frac{pL}{8G}\Bigg\{&(Z-\bar{Z})\left[\frac{Z}{\sqrt{(Z+1)(Z-1)}} - 1\right] + \kappa\left[\sqrt{(\bar{Z}+1)(\bar{Z}-1)} - \bar{Z}\right]\\ &-\left[\sqrt{(Z+1)(Z-1)} - Z\right]\Bigg\}.\end{aligned} \tag{9.243}$$

The tractions are given by (9.183) on page 142 with the expressions (9.239) for τ^{11} and τ^{12}, with $\alpha = 0$,

$$t_s - \mathrm{i}t_n = T^1 = -\frac{1}{2}\mathrm{i}(\tau^{11} - \tau^{12}) = -\tfrac{1}{2}\mathrm{i}\left\{-\frac{L}{2}(Z-\bar{Z})\Phi''(z) + \Phi'(z) + \bar{\Phi}(\bar{z})\right\}. \quad (9.244)$$

We differentiate the function Φ', (9.242), with respect to z,

$$\Phi''(z) = \frac{2p}{L}\left[\frac{1}{\sqrt{(Z+1)(Z-1)}} - \frac{Z^2}{(Z^2-1)\sqrt{(Z+1)(Z-1)}}\right], \quad (9.245)$$

and combine terms,

$$\Phi''(z) = \frac{2p}{L}\frac{1}{\sqrt{(Z-1)(Z+1)}}\left[\frac{Z^2-1-Z^2}{Z^2-1}\right] = -\frac{2p}{L}\frac{1}{(Z^2-1)\sqrt{(Z-1)(Z+1)}}. \quad (9.246)$$

The Displacements along the Crack
Along the crack, $Z - \bar{Z} = 0$, so that only the two trailing terms in (9.243) contribute. The argument of $Z-1$ is π along the plus side ($Y = 0^+$) of the crack, and $-\pi$ along the minus side; the argument of $\bar{Z}-1$ is $-\pi$ along the plus side, and π along the minus side. We use this in (9.243):

$$\begin{aligned} w^+ &= \frac{pL}{8G}\left[-\mathrm{i}\kappa\sqrt{1-X^2} - \kappa X - \mathrm{i}\sqrt{1-X^2} + X\right] \\ &= -\frac{pL}{8G}\left[\mathrm{i}(1+\kappa)\sqrt{1-X^2} + (\kappa-1)X\right]. \end{aligned} \quad (9.247)$$

We obtain a similar expression for w^-,

$$\begin{aligned} w^- &= \frac{pL}{8G}\left[\mathrm{i}\kappa\sqrt{1-X^2} - \kappa X + \mathrm{i}\sqrt{1-X^2} + X\right] \\ &= -\frac{pL}{8G}\left[-\mathrm{i}(1+\kappa)\sqrt{1-X^2} + (\kappa-1)X\right]. \end{aligned} \quad (9.248)$$

The jump in displacement is purely imaginary,

$$[w] = -\mathrm{i}[u_n] = -\mathrm{i}\frac{pL(1+\kappa)}{4G}\sqrt{1-X^2}. \quad (9.249)$$

A single symbol inside brackets, as in $[w]$, indicates a jump. The jump in normal displacement is positive, i.e., the normal displacement, which is positive in the direction of the Y-axis, is larger on the $+$ side than on the $-$ side; the crack opens. A plot of normal and tangential components of the displacements along the crack are shown in Figure 9.27. A vector plot of the displacement around the crack is shown in Figure 9.28.

Tractions along the Crack
The tractions are given by (9.244), with $Z - \bar{Z} = 0$ along the crack,

$$t_s - \mathrm{i}t_n = T^1 = -\tfrac{1}{2}\mathrm{i}\left\{\Phi'(z) + \bar{\Phi}'(\bar{z})\right\} \qquad Z = \bar{Z}. \quad (9.250)$$

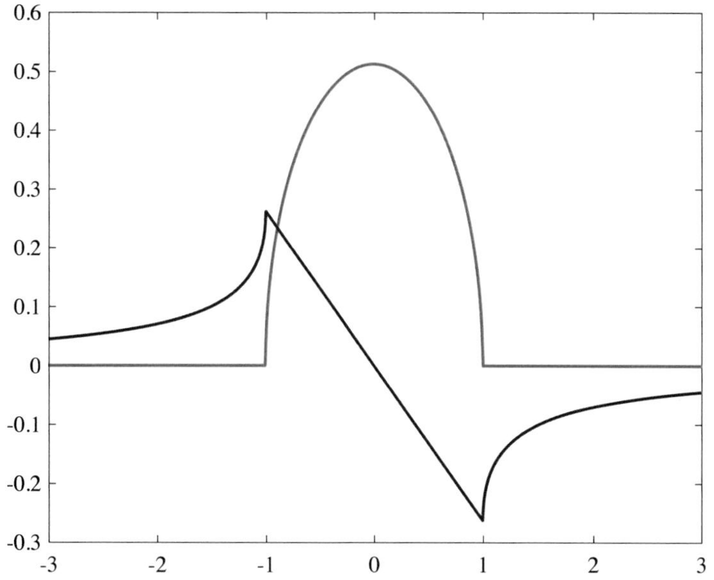

Figure 9.27: Tangential (blue) and normal (red) displacements along the crack

We use expression (9.242) for $\Phi'(z)$:

$$t_s - \mathrm{i}t_n = -\frac{\mathrm{i}}{2}\left\{\Phi'(z) + \bar{\Phi}'(\bar{z})\right\} = -\frac{\mathrm{i}p}{2}\left\{\frac{Z}{\sqrt{(Z-1)(Z+1)}} - 1 + \frac{\bar{Z}}{\sqrt{(\bar{Z}-1)(\bar{Z}+1)}} - 1\right\}. \tag{9.251}$$

We verify that the tractions are continuous across the crack, and obtain, for the plus side of the crack:

$$(t_s - \mathrm{i}t_n)^+ = -\frac{\mathrm{i}p}{2}\left\{\frac{X}{\mathrm{i}\sqrt{1-X^2}} + \frac{X}{-\mathrm{i}\sqrt{1-X^2}} - 2\right\} = \mathrm{i}p. \tag{9.252}$$

We apply a similar approach to the - side

$$(t_s - \mathrm{i}t_n)^- = \mathrm{i}p. \tag{9.253}$$

The shear traction is zero, as required, and the normal traction is constant and points into the crack; it is equal to $-p$ on both sides of the crack.

A plot of the major principal stresses is shown in Figure 9.29.

Tractions along the Real Axis outside the Crack
When $X < -1, Y = 0$, the arguments of $Z+1$ and $Z-1$ are both equal to π, so that the square roots in (9.252) are real; we obtain:

$$t_s - \mathrm{i}t_n = \mathrm{i}p\left[\frac{X}{\sqrt{1-X^2}} + 1\right]. \tag{9.254}$$

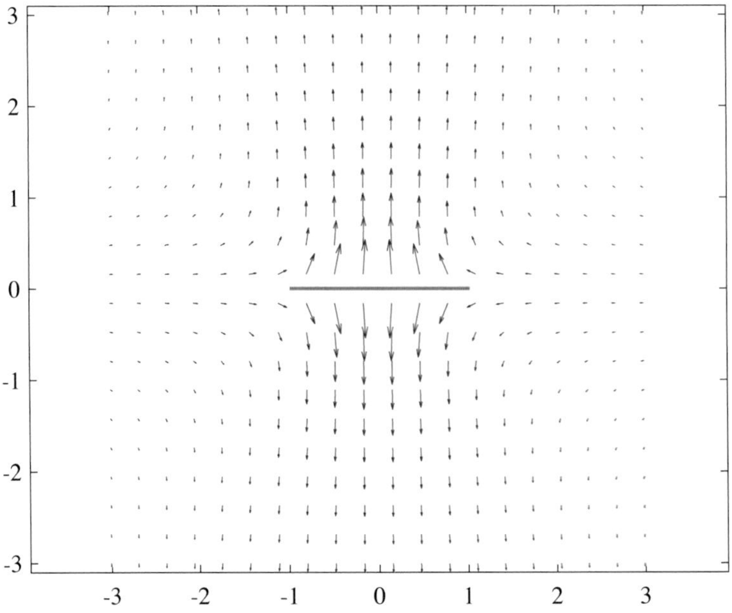

Figure 9.28: Displacement field for a pressurized crack

When $X > 1, Y = 0$, (9.251) becomes:

$$t_s - \mathrm{i}t_n = -\mathrm{i}p\left\{\frac{X}{\sqrt{X^2-1}} - 1\right\}. \tag{9.255}$$

The normal stresses are singular at the crack tips.

Behavior at Infinity
We expand the displacement function, w, (9.238), about infinity to examine the far-field behavior of w. We expand $\Phi'(z)$ about infinity:

$$\Phi'(z) = \lim_{Z\to\infty} p\left[\frac{Z}{\sqrt{Z^2-1}} - 1\right] = \lim_{Z\to\infty} p\left[\left(\frac{1}{1-Z^{-2}}\right)^{1/2} - 1\right], \tag{9.256}$$

or

$$\Phi'(z) = p\left(1 + \frac{1}{2Z^2} + \cdots - 1\right) = p\left(\frac{1}{2Z^2} + \cdots\right). \tag{9.257}$$

We also expand $\Phi(z)$, (9.240), about infinity:

$$\Phi = \frac{pL}{2}Z\left[\sqrt{1-\frac{1}{Z^2}} - 1\right] = \frac{pL}{2}Z\left[1 - \frac{1}{2Z^2} + \cdots - 1\right] = \frac{pL}{2}\frac{1}{2Z} + \cdots. \tag{9.258}$$

All terms in the expression for the displacement, (9.243), vanish near infinity.

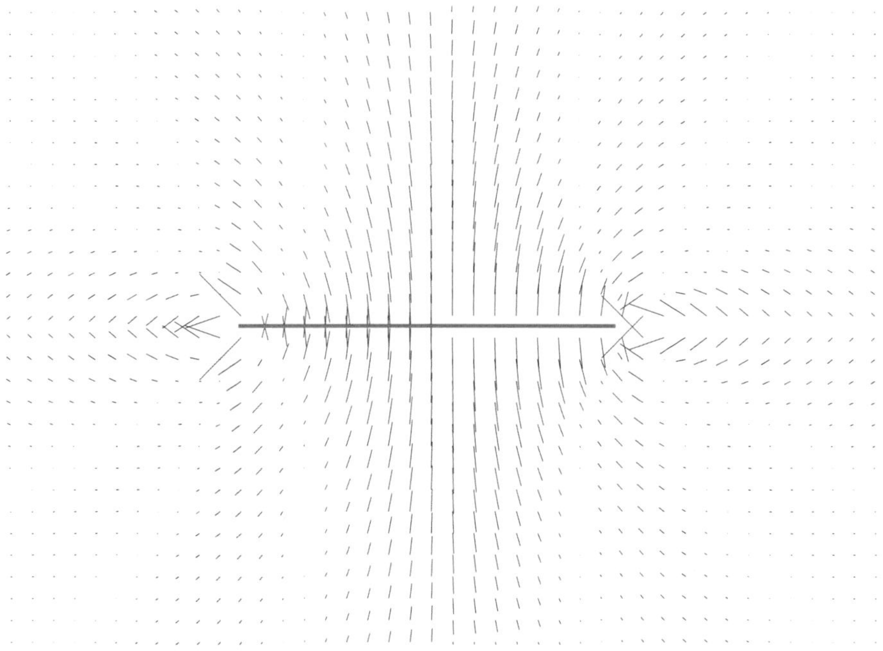

Figure 9.29: Major principal stresses for a crack, shown in red in the figure. The lengths of the lines are proportional to the magnitude of the stress

Problem 9.7

Consider a crack extending from $z = -100$ to $z = 100$, with a pressure equal to p.

Questions:

1. Present expressions for the three components of the stress tensor in Cartesian space.
2. Expand all components of the stress tensor about infinity.

9.5.10 An Analytic Element for Gravity in a Half-Space

The purpose of the analytic element for gravity is to include the effect of gravity by superposition, so that the gravity terms may be omitted from all other elements. We create the element such that the x-axis is stress-free. The displacements and tractions along the half space are given by (9.173), (9.174), and (9.175) on page 141,

$$w = \frac{1}{4G}\left\{ -\bar{z}\Phi'(z) + \kappa\bar{\Phi}(\bar{z}) + \Psi(z) - 2(1-2\nu)\bar{B}_2 \right\}, \tag{9.259}$$

and

$$\tau^{11} = \bar{z}\Phi''(z) - \Psi'(z), \tag{9.260}$$

and

$$\tau^{12} = -\Phi'(z) - \bar{\Phi}'(\bar{z}) + \left(\bar{B}_1 + B_1\right). \tag{9.261}$$

We choose $\Psi(z)$ as:

$$\Psi(z) = z\Phi'(z) + \psi(z), \tag{9.262}$$

so that expression (9.259) for w becomes

$$w = \frac{1}{4G}\{(z-\bar{z})\Phi'(z) + \kappa\bar{\Phi}(\bar{z}) + \psi(z) - 2(1-2\nu)\bar{B}_2\}. \tag{9.263}$$

We differentiate (9.262),

$$\Psi'(z) = \Phi'(z) + z\Phi''(z) + \psi'(z), \tag{9.264}$$

and use this in the expressions for the components of the stress tensor:

$$\tau^{11} = -(z-\bar{z})\Phi''(z) - \Phi'(z) - \psi'(z), \tag{9.265}$$

and

$$\tau^{12} = -\Phi'(z) - \bar{\Phi}'(\bar{z}) + (\bar{B}_1 + B_1). \tag{9.266}$$

We express the tractions along $z = \bar{z}$ by the use of (9.183) on page 142, with $\alpha = 0$,

$$t_1 - \mathrm{i}t_2 = -\tfrac{1}{2}\mathrm{i}[\tau^{11} - \tau^{12}] = -\tfrac{1}{2}\mathrm{i}[-(z-\bar{z})\Phi''(z) - \psi'(z) + \bar{\Phi}'(\bar{z}) - B_1 - \bar{B}_1]. \tag{9.267}$$

The functions B_1 and B_2 are given by (9.230) and the expression for w becomes

$$w = \frac{1}{4G}\{(z-\bar{z})\Phi'(z) + \kappa\bar{\Phi}(\bar{z}) + \psi(z) + \rho g\mathrm{i}(1-2\nu)\bar{z}^2\}. \tag{9.268}$$

We choose particular solutions for Φ and ψ to account for the effect of gravity

$$\Phi(z) = \mathrm{i}\rho g\frac{1-2\nu}{\kappa+1}z^2 \qquad \psi = -\mathrm{i}\rho g\frac{1-2\nu}{\kappa+1}z^2. \tag{9.269}$$

We use these expressions in (9.264) for $\Psi'(z)$

$$\Psi'(z) = 2\mathrm{i}\rho g\frac{1-2\nu}{\kappa+1}[z+z-z] = 2\mathrm{i}\rho g\frac{1-2\nu}{\kappa+1}z, \tag{9.270}$$

and use this with the expression for Φ in (9.260) and (9.261),

$$\boxed{\tau^{11} = -2\mathrm{i}\rho g\frac{1-2\nu}{\kappa+1}(z-\bar{z})}, \tag{9.271}$$

and

$$\tau^{12} = -2\mathrm{i}\rho g\frac{1-2\nu}{\kappa-1}(z-\bar{z}) + \mathrm{i}\rho g(z-\bar{z}) = \mathrm{i}\rho g\frac{1}{\kappa+1}[-2+4\nu+\kappa+1](z-\bar{z}), \tag{9.272}$$

or, with $\kappa = 3-4\nu$

$$\boxed{\tau^{12} = 2\mathrm{i}\rho g\frac{1}{\kappa+1}(z-\bar{z})}. \tag{9.273}$$

Both components of the transformed stress tensor vanish along the upper boundary of the half space, $z = \bar{z}$, so that the element does not contribute to the tractions there. We follow a similar process to obtain an expression for the displacements, (9.268),

$$\begin{aligned} w &= \mathrm{i}\rho g \frac{1-2\nu}{\kappa+1}\frac{1}{4G}\left[2(z-\bar{z})z - \kappa\bar{z}^2 - z^2 + (\kappa+1)\bar{z}^2\right] \\ &= \mathrm{i}\rho g \frac{1-2\nu}{\kappa+1}\frac{1}{4G}\left[z^2 - 2z\bar{z} + \bar{z}^2\right], \end{aligned} \tag{9.274}$$

or

$$\boxed{w = \mathrm{i}\rho g \frac{1-2\nu}{\kappa+1}\frac{1}{4G}(z-\bar{z})^2}. \tag{9.275}$$

The displacements along the surface $z = \bar{z}$ are zero.

Expressions for the Stress Tensor and Displacement Components

We find expressions for the stresses from

$$\sigma_{xx} = \tfrac{1}{2}(\tau^{11} + \tau^{22}) = -\frac{2\mathrm{i}\rho g}{\kappa+1}(1-2\nu-1)(z-\bar{z}) = \mathrm{i}\rho g \frac{2\nu}{4(1-\nu)}2\mathrm{i}y \tag{9.276}$$

$$\sigma_{yy} = -\tfrac{1}{2}(\tau^{11} - \tau^{22}) = \frac{\mathrm{i}\rho g}{\kappa+1}(1-2\nu+1)(z-\bar{z}) = \mathrm{i}\rho g \frac{2(1-\nu)}{4(1-\nu})2\mathrm{i}y, \tag{9.277}$$

where we used that $\kappa + 1 = 4(1-\nu)$. We simplify:

$$\sigma_{xx} = -\rho g \frac{\nu}{1-\nu}y \tag{9.278}$$

$$\sigma_{yy} = -\rho g y. \tag{9.279}$$

Both stress tensor components are positive, and thus compressive, (y is negative), and increase linearly with depth. The vertical component increases as $\rho g|y|$, as expected. The horizontal stress is generally unknown, but for this solution is compressive and also increases linearly with depth.

We separate real and imaginary parts in (9.275),

$$w = u_x - \mathrm{i}u_y = -\rho g \mathrm{i}\frac{1-2\nu}{\kappa+1}\frac{1}{G}y^2, \tag{9.280}$$

so that

$$u_x = 0 \tag{9.281}$$

$$u_y = \frac{\rho g}{G}\frac{1-2\nu}{\kappa+1}(y^2 - H^2). \tag{9.282}$$

We see that u_x is zero, and we added a constant to u_y, chosen such that the displacement is zero at $y = -H$. The displacement u_y increases without limit with depth, a result of gravity acting on a domain of infinite depth. We usually deal with this problem by setting the displacement to zero at some distance from the surface.

Problem 9.8

Use the stress-strain relations and the expressions for the stresses to show that the strain e_{xx} vanishes everywhere in the lower half plane.

Chapter 10

The Parabolic Case: Two Coinciding Characteristics

Parabolic partial differential equations have two coinciding characteristics, which makes them less suitable for the application of complex variables, in contrast to elliptical partial differential equations. A partial differential equation that is common in engineering practice is the *diffusion equation*, also known as the *heat equation*; it occurs in problems of diffusion as well as in problems of heat conduction in solids. This differential equation is not limited to these applications; it also describes consolidation of soils and problems of transient groundwater flow.

The objective of this chapter is to demonstrate the most common approaches for solving this partial differential equation: the application of Laplace transforms and the method of separation of variables. We begin by deriving the diffusion equation, followed by a brief discussion of the Laplace transform. We apply the Laplace transform to a few problems of consolidation of soils, and finally use the method of separation of variables to solve a problem of groundwater flow. These applications could equally well be modified to apply to problems of diffusion or heat flow.

We focus our attention on applications to illustrate how Laplace transforms can be used to solve practical problems, and for more information refer the readers to the extensive literature on operational mathematics, e.g., Churchill [1958].

10.1 The Diffusion Equation

Diffusion is a mechanism that describes, among other things, the spreading of some material (for example a pollutant) in a liquid. The underlying constitutive equation is *Fick's law*, which states that the transport of some material out of an elementary volume is proportional to minus the gradient of the concentration (mass per unit volume of fluid: [M/L^3]). The mass transport is in the direction of decreasing concentration, c. If the mass flux in the i-direction is F_i [M/(L^{2}T)], then Fick's law is:

$$F_i = -\alpha \frac{\partial c}{\partial x_i}, \tag{10.1}$$

O. D. L. Strack, *Applications of Vector Analysis and Complex Variables in Engineering*, https://doi.org/10.1007/978-3-030-41168-8_10

where α is a proportionality constant, the diffusion coefficient, with dimension [L^2/T]. The net flow out of an elementary unit volume is equal to the divergence of F_i, $\partial F_i/\partial x_i$, and must be equal, by mass balance, to the decrease in mass per unit volume and time, $-\partial c/\partial t$. Hence,

$$\frac{\partial F_i}{\partial x_i} = -\frac{\partial c}{\partial t}, \tag{10.2}$$

or, with (10.1), for constant α

$$-\alpha \frac{\partial^2 c}{\partial x_i \partial x_i} = -\frac{\partial c}{\partial t}, \tag{10.3}$$

so that

$$\nabla^2 c = \frac{\partial^2 c}{\partial x_i \partial x_i} = \frac{1}{\alpha}\frac{\partial c}{\partial t}. \tag{10.4}$$

10.1.1 Characteristics

We showed in section 7.3 that the diffusion equation is parabolic, and that the single characteristic implies that discontinuities, such as an instantaneous change in the dependent variable, the concentration for example, has an immediate effect all the way to infinity because the characteristic is a horizontal line in an x, τ-space.

The diffusion equation is parabolic also for two- and three- dimensional cases; the type of partial differential equation is independent of the number of spatial dimensions involved.

We discuss in what follows the flow of heat in solids, one-dimensional consolidation of soils, and three-dimensional transient groundwater flow. These processes are characterized by equations of motion similar to Fick's law.

10.2 Conduction of Heat in Solids

The equation for the flux of heat through a solid, f_i, is

$$f_i = -K\frac{\partial T}{\partial x_i}, \tag{10.5}$$

where T is the temperature, and K is the conductivity of the solid. The parabolic partial differential equation that governs the conduction of heat in solids is, see Carslaw and Jaeger [1959],

$$\nabla^2 T = \frac{1}{\kappa}\frac{\partial T}{\partial t}, \tag{10.6}$$

where κ is the diffusivity,

$$\kappa = \frac{K}{\rho c}, \tag{10.7}$$

ρ is the density of the solid, and c its specific heat.

10.3 Transient Groundwater Flow

We briefly discussed groundwater flow in section 9.4 for steady flow. The equation of motion is Darcy's law,

$$q_i = -k\frac{\partial \phi}{\partial x_i} \qquad i = 1,2,3 \tag{10.8}$$

where q_i is the specific discharge [L/T], representing the discharge flowing through a unit area with its normal in the direction of the component, where k [L/T] is the hydraulic conductivity, and ϕ is the hydraulic head [L]. We consider the approximate equation of motion for vertically integrated flow, see e.g., Strack [2017],

$$Q_i = -\frac{\partial \Phi}{\partial x_i} \qquad i = 1,2. \tag{10.9}$$

The discharge potential Φ is defined as

$$\Phi = kH\phi \tag{10.10}$$

where H is the aquifer thickness, and Q_i is the discharge vector, representing the total flow passing through a vertical plane covering the entire thickness of the aquifer, of unit width, and with normal in the direction of the component. Equation (10.9) is based on the Dupuit-Forchheimer approximation [Dupuit, 1863, Forchheimer, 1886], which implies that the horizontal components of the specific discharge vector do not vary vertically.

For transient flow, the divergence of the discharge vector Q_i is equal to a factor, the specific storage coefficient S_s [L^{-1}], divided by the hydraulic conductivity, multiplied by the rate of change of the potential with respect to time. Combining this with Darcy's law, we obtain the governing equation for transient vertically integrated groundwater flow,

$$\nabla^2\Phi = \frac{S_s}{k}\frac{\partial \Phi}{\partial t}. \tag{10.11}$$

10.4 Consolidation of Soils

We consider the consolidation of a body of soil, usually clay, that is compressed in the vertical direction, for example in response to the application of load. The process is quite complex; the soil is composed of grains, forming a skeleton, with water filling the pore space between the grains. Often the compressibility of the water is small compared to that of the grain skeleton; we neglect this for the following examples. Deformation of the soil is then possible only upon removal of the water, i.e., it requires flow of water through the soil, which takes time. The result is a behavior that can be surprising, as we see in the following examples.

The theory of one-dimensional consolidation is based on a concept proposed by Karl von Terzaghi, an Austrian engineer who was the original developer of soil mechanics, see Terzaghi [1925]. He proposed to define total and effective stresses, taken positive for compression, see e.g.,Scott [1963], Verruijt [2017]. We consider displacements only in the vertical direction and deal with the total stress σ_{zz} and the effective stress σ'_{zz}, related to the pore pressure p as

$$\sigma_{zz} = \sigma'_{zz} + p. \tag{10.12}$$

The effective stresses are an approximation of the stresses between the grains. We consider only the responses in stresses, strains, and pressures due to a *change* in conditions, and add the superscript c to remind us of this property; (10.12) becomes

$$\overset{c}{\sigma}_{zz} = \overset{c}{\sigma}{}'_{zz} + \overset{c}{p}. \tag{10.13}$$

The volume strain e_0 is equal to e_{zz} for this case of one-dimensional consolidation, and conservation of mass requires that[1]

$$\frac{\partial e_{zz}}{\partial t} = \frac{\partial \overset{c}{q}_z}{\partial z} \tag{10.14}$$

We apply Darcy's law to the change in flow, represent the unit weight of water as γ_w, and use that $\overset{c}{\phi} = \overset{c}{p}/\gamma_w$ because the elevation head does not change,

$$\overset{c}{q}_z = -k\frac{\partial \overset{c}{\phi}}{\partial z} = -\frac{k}{\gamma_w}\frac{\partial \overset{c}{p}}{\partial z}. \tag{10.15}$$

We use this with (10.14),

$$\frac{\partial e_{zz}}{\partial t} = -\frac{k}{\gamma_w}\frac{\partial^2 \overset{c}{p}}{\partial z^2} \tag{10.16}$$

The strain e_{zz} is related to the change in effective stress via a constant, m_v [L^2/F], the coefficient of volume compressibility, determined in a one-dimensional compression test,

$$e_{zz} = m_v \overset{c}{\sigma}{}'_{zz} = m_v\left(\overset{c}{\sigma}_{zz} - \overset{c}{p}\right) \tag{10.17}$$

We introduce a new variable, P,

$$P = \overset{c}{p} - \overset{c}{\sigma}_{zz}, \tag{10.18}$$

divide both sides of (10.16) by m_v, use expression (10.17) for e_{zz}, and replace $\overset{c}{p}$ by P, restricting $\overset{c}{\sigma}_{zz}$ to being independent of z,

$$c_v\frac{\partial^2 P}{\partial z^2} = \frac{\partial P}{\partial t} \tag{10.19}$$

where c_v [L^2/T] is the consolidation coefficient,

$$c_v = \frac{k}{m_v \gamma_w}. \tag{10.20}$$

[1] We neglect the deformation of the grains, so that the volume strain represents the decrease in pore space.

10.5 The Laplace Transform

The Laplace transform is the transformation of a function of a certain variable, usually time (t), into another function of another variable, usually s. There is an extensive body of literature available on operational mathematics, including the theory and application of the Laplace transform; we present the method here in sufficient detail for application to several problems taken from engineering practice. The reader is referred to the literature for a more detailed description of the method, as well as rigorous derivations of the equations used here.

The Laplace transform of a function $f(t)$ is defined as

$$\mathscr{L}(f) = \tilde{f}(s) = \int_0^{\infty} f(t)\mathrm{e}^{-st}dt, \tag{10.21}$$

where $\mathscr{L}(f)$ is the Laplace transform of the function $f(t)$. We represent the Laplace transform of a function by placing a tilde over the symbol, i.e., $\tilde{f}(s)$ is the Laplace transform of $f(t)$.

The benefit of the Laplace transform is that the transform of the derivative of $f(t)$ with respect to t is s times the transformed function plus the initial value of the original function; the operation of differentiation is transformed into multiplication by the independent variable. That this is true can be verified,

$$\mathscr{L}(f'(t)) = \int_0^{\infty} f'(t)\mathrm{e}^{-st}dt = \int_0^{\infty} \mathrm{e}^{-st}df = \mathrm{e}^{-st}f(t)\Big|_0^{\infty} - \int_0^{\infty} f(t)(-s)\mathrm{e}^{-st}dt, \tag{10.22}$$

where the prime denotes differentiation with respect to time. We restrict the application of the Laplace transform to cases where

$$\lim_{t\to\infty} \mathrm{e}^{-st}f(t) = 0, \tag{10.23}$$

so that (10.22) reduces to

$$\mathscr{L}(f'(t)) = s\int_0^{\infty} f(t)dt - f(0) = sf(s) - f(0) \tag{10.24}$$

We apply the Laplace transform to the heat equation (10.6),

$$\nabla^2\tilde{T} = \frac{1}{\kappa}[s\tilde{T}(s) - T(0)], \tag{10.25}$$

where $\tilde{T}(s)$ is the Laplace transform of $T(t)$. Differentiation with respect to the spatial coordinates is not affected by the transform; we reduced the original partial differential equation in terms of space and time into a new one, in terms of space only.

After solving the problem in the transformed domain, the Laplace domain, we must apply the inverse process, the inverse Laplace transform, in order to obtain the solution in terms of space and time. This can often be done simply by looking up tabulated inverse transforms, by formally applying the inverse transformation process using complex contour integration, not discussed here, or by using a numerical inverse transform.

10.5.1 The Unit Step Function

A property of Laplace transforms is that the lower bound of the integral is $t = 0$. To understand the consequence of this, we consider the function

$$\begin{aligned} U(t) &= 0 \qquad t \leq 0 \\ U(t) &= 1 \qquad t > 0 \end{aligned} \tag{10.26}$$

This function is known as the Heaviside unit step function, usually represented as $U(t)$, and is illustrated in Figure 10.1. We obtain the Laplace transform of this function from

$$\tilde{U} = \int_0^\infty \mathrm{e}^{-st} dt = -\frac{1}{s} \int_0^\infty d\mathrm{e}^{-st} = \frac{1}{s} \tag{10.27}$$

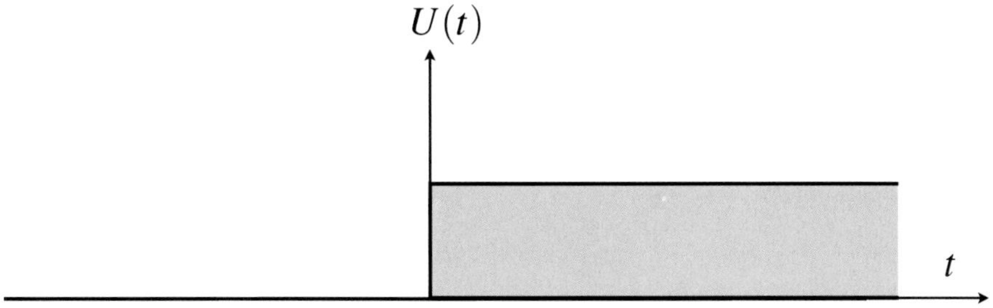

Figure 10.1: The unit step function, $U(t)$.

10.5.2 The Dirac Delta Function

The Dirac delta function $\delta(t)$ is the derivative of the unit step function; this function is infinite at $t = 0$, is zero everywhere else, and has the property that the integral of the delta function over an interval that includes $t = 0$ is equal to 1. We obtain the Laplace transform of the Dirac delta function from the notion that it is the derivative of the unit step function, i.e.,

$$\tilde{\delta}(s) = sU - U(0) = 1, \tag{10.28}$$

where we take the initial value of U as zero, according to the definition (10.26).

Complex Representation of U and δ

It is possible to represent the unit step function as the real part of a complex function:

$$U = \Re \frac{1}{\mathrm{i}\pi} [\mathrm{i}\pi - \ln(z)] \qquad \Im z = 0^+ \tag{10.29}$$

When z is real, with its argument equal to 0 for $x \geq 0$ the function U equals 1, and when z is real with its argument equal to π for $x < 0$, U equals 0.

The function U as defined in (10.29) is differentiable, and we obtain the Dirac delta function by differentiation with respect to z,

$$\delta = \frac{dU}{dz} = -\Re\frac{1}{\mathrm{i}\pi}\frac{1}{z} \qquad \Im z = 0^{+} \tag{10.30}$$

This representation of the Dirac delta function meets the condition that its integral over the entire real axis is 1.

10.6 Applications of the Laplace transform

We apply the Laplace transform to solve several problems of one-dimensional consolidation, to determine the solution for heat flow, or groundwater flow, due to a sink applied for a one-dimensional problem of heat or groundwater flow induced by a vertical rod in an infinite slab.

10.6.1 One-Dimensional Consolidation

We cover three applications of one-dimensional consolidation using Laplace transforms. We transform the function P, defined as $\overset{c}{p} - \overset{c}{\sigma}_{zz}$, to $\tilde{P}$ and transform the differential equation (10.19):

$$c_v\frac{\partial^2\tilde{P}}{\partial z^2} = s\tilde{P} - P(0). \tag{10.31}$$

An applied load is initially taken up entirely by the pore pressure; we consider the fluid incompressible and changing the pore space requires fluid flow, which takes time. If we do not apply a load, then the initial pore pressure change, $\overset{c}{p}$, is zero. In either case, the initial value of P, $P(0)$, is zero, so that (10.31) reduces to

$$c_v\frac{\partial^2\tilde{P}}{\partial z^2} = s\tilde{P}. \tag{10.32}$$

A solution to this equation is the exponential function,

$$\tilde{P} = A\mathrm{e}^{a(z+\beta)}, \tag{10.33}$$

where A, α, and β are constants[2], and substitution in the governing equation gives

$$\tilde{P} = c_v a^2 A\mathrm{e}^{a(z+\beta)} = sA\mathrm{e}^{a(z+\beta)}, \tag{10.34}$$

We divide by the common factor, and find the following expression for a,

$$a = \pm\sqrt{\frac{s}{c_v}} = \pm\alpha, \tag{10.35}$$

[2] A and β together represent a single constant, as adding a constant to the argument of an exponential function amounts to multiplication the exponential by a constant.

where

$$\alpha = \sqrt{\frac{s}{c_v}} \tag{10.36}$$

The general solution for $\tilde{P}$ is

$$\tilde{P} = A\mathrm{e}^{\alpha(z+\beta)} + B\mathrm{e}^{-\alpha(z+\beta)}. \tag{10.37}$$

Compression of a Clay Layer Due to Loading

Consider the problem of consolidation of a layer of clay between two layers of highly permeable sand. A load is applied over a very short period of time on top of the upper sand layer, and then kept in place, see Figure 10.2. The load, $\overset{c}{\sigma}_{zz} = \overset{0}{p}$, is applied at time

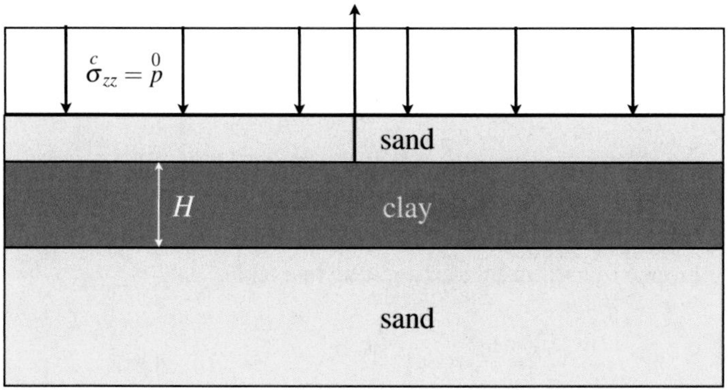

Figure 10.2: Consolidation of a layer of clay between two sand layers, loaded by a surcharge.

$t = 0$. We choose the origin of the coordinate system at the upper boundary of the clay layer of thickness H, see Figure 10.2. We consider the sand layers infinitely permeable as compared to the clay layer, so that the pressure change is zero at these surfaces,

$$\begin{aligned} &\overset{c}{p}(0,t) = 0 \qquad P(0,t) = \overset{c}{p}(0,t) - \overset{0}{p} = -\overset{0}{p} \\ &\overset{c}{p}(-H,t) = 0 \quad P(-H,t) = \overset{c}{p}(0,t) - \overset{0}{p} = -\overset{0}{p}. \end{aligned} \tag{10.38}$$

We transform the boundary conditions (10.38),

$$\begin{aligned} \tilde{P} &= -\frac{\overset{0}{p}}{s} \qquad z = 0 \\ \tilde{P} &= -\frac{\overset{0}{p}}{s} \qquad z = -H. \end{aligned} \tag{10.39}$$

The boundary values of $\tilde{P}$ are the same at the upper and lower boundaries of the clay layer; the solution is symmetrical with respect to the line $z = -H/2$; it is advantageous to introduce a new dimensionless variable Z as

$$Z = \frac{z}{H} + \frac{1}{2}, \tag{10.40}$$

i.e., we have chosen β in (10.33) as $H/2$. The function $\tilde{P}$ is symmetrical with respect to $Z = 0$, and has the form

$$\tilde{P} = C\cosh(\alpha H Z). \tag{10.41}$$

The boundary condition is that $\tilde{P}$ is $-\overset{0}{p}/s$ at $Z = \pm 1/2$,

$$C\cosh(\tfrac{1}{2}\alpha H) = -\frac{\overset{0}{p}}{s} \to C = -\frac{\overset{0}{p}}{s\cosh(\frac{1}{2}\alpha H)}, \tag{10.42}$$

and the expression for $\tilde{P}$ becomes

$$\tilde{P} = -\overset{0}{p}\frac{\cosh(\alpha H Z)}{s\cosh(\frac{1}{2}\alpha H)}. \tag{10.43}$$

or, with $\alpha = \sqrt{s/c_v}$,

$$\tilde{P} = -\overset{0}{p}\frac{\cosh(HZ\sqrt{s/c_v})}{s\cosh(\frac{1}{2}H\sqrt{s/c_v})}. \tag{10.44}$$

We find the inverse transform by looking it up in a table,

$$P = -\overset{0}{p} - \frac{4\overset{0}{p}}{\pi}\sum_{n=1}^{\infty}\frac{(-1)^n}{2n-1}\mathrm{e}^{-(2n-1)^2\pi^2 t c_v/H^2}\cos\left[(2n-1)\pi Z\right]. \tag{10.45}$$

We simplify this, and introduce dimensionless time τ as

$$\tau = \frac{t c_v}{H^2}. \tag{10.46}$$

Recall the definition of P;

$$P = \overset{c}{p} - \overset{0}{p}, \tag{10.47}$$

which we solve for $\overset{c}{p}$ with (10.45):

$$\overset{c}{p} = -\frac{4\overset{0}{p}}{\pi}\sum_{n=1}^{\infty}\frac{(-1)^n}{2n-1}\mathrm{e}^{-(2n-1)^2\pi^2\tau}\cos[(2n-1)\pi Z]. \tag{10.48}$$

Curves of $\overset{c}{p}/\overset{0}{p}$ versus Z for constant values of τ are shown in Figure 10.3. At time $\tau = 0$, the pressure takes up the entire load, i.e., $\overset{c}{p}$ equals $\overset{0}{p}$. The pressures at the boundaries $Z = \pm 1/2$ are kept to zero, and the excess pore pressure $\overset{c}{p}$ reduces to zero for $t \to \infty$.

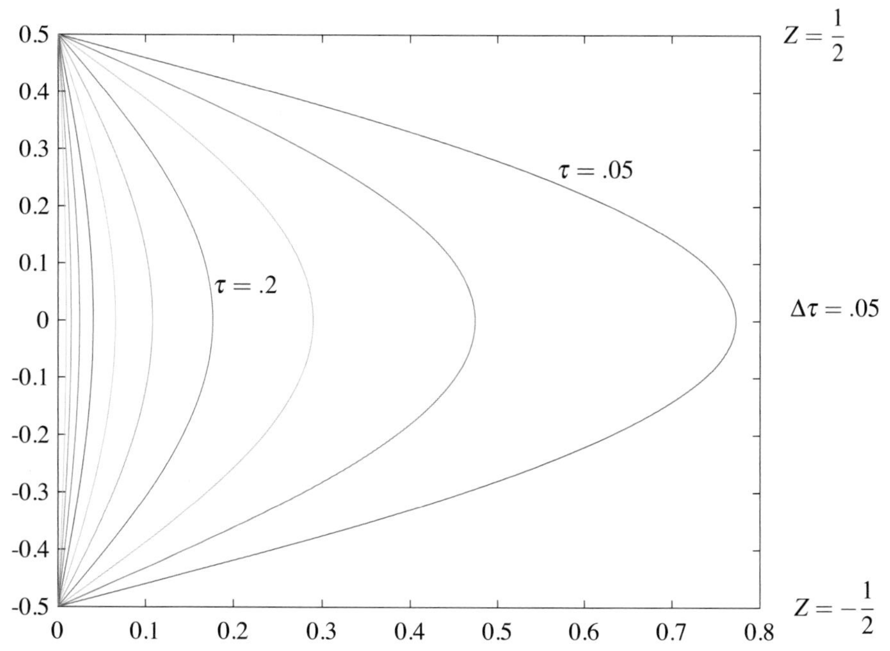

Figure 10.3: Curves of $\overset{c}{p}/\overset{0}{p}$ versus Z for constant values of time. The difference in value of τ between curves is 0.05. and the value of τ for the leftmost curve is .5. The initial value of $\overset{c}{p}/\overset{0}{p}$ is 1

One-Dimensional Compression Test

Consider the deformation of a soil sample subjected to a load in a one-dimensional compression test; a cylindrical sample is subjected to a load transmitted via a very stiff top plate. The cylinder is so stiff that its deformation can be neglected. The base plate is impermeable. The testing apparatus is illustrated in Figure 10.4. The problem is similar to the previous application, except for the condition along the base plate, which is a no-flow boundary. We use the same variables as in the preceding example. The initial pressure is equal to the load. The top plate is made of porous material, so that the pressure change is zero there. This feature of the apparatus is necessary; if both upper and lower boundaries were impermeable, there would be deformation only if the water were compressible. The initial condition is the same as before, but the condition along the lower boundary is:

$$\frac{\partial \overset{c}{p}}{\partial z} = 0 \quad \rightarrow \quad \frac{\partial P}{\partial z} = 0. \tag{10.49}$$

The general solution has the same form as before. We choose β as H; the expression for $\tilde{P}$ is:

$$\tilde{P} = A\mathrm{e}^{\alpha(z+H)} + B\mathrm{e}^{-\alpha(z+H)}, \tag{10.50}$$

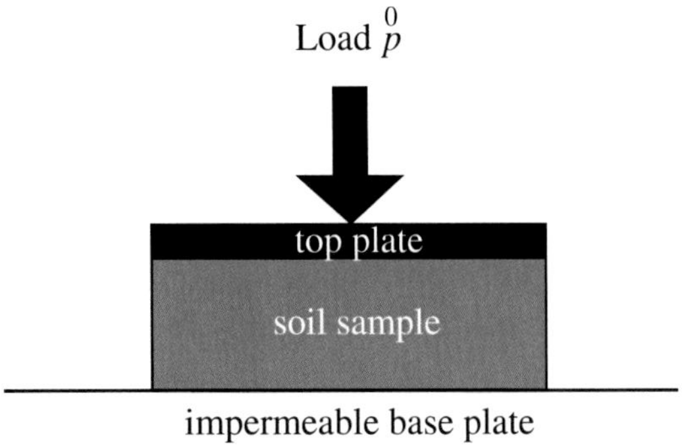

Figure 10.4: One-dimensional compression test

and we apply the boundary conditions:

$$\begin{aligned} \tilde{P} &= -\frac{\overset{0}{p}}{s} = Ae^{\alpha H} + Be^{-\alpha H} \qquad && z = 0 \\ \frac{\partial \tilde{P}}{\partial z} &= 0 = A\alpha - B\alpha && z = -H, \end{aligned} \tag{10.51}$$

The second equation gives $A = B$, and the first one yields

$$-\frac{\overset{0}{p}}{s} = 2A\cosh(\alpha H) \rightarrow A = -\frac{\overset{0}{p}/s}{2\cosh(\alpha H)}. \tag{10.52}$$

We use this in (10.50),

$$\tilde{P} = -\frac{\overset{0}{p}}{s}\frac{\cosh\{\alpha(z+H)\}}{\cosh(\alpha H)} = -\frac{\overset{0}{p}}{s}\frac{\cosh\{\sqrt{s/c_v}(z+H)\}}{\cosh(\sqrt{s/c_v}H)}. \tag{10.53}$$

This solution is similar to that obtained for the first case. The main difference is that the argument of the cosine in the denominator does not have the factor 1/2. The solutions are similar because the horizontal plane through the center of the clay layer acts as an impermeable boundary; the upper half of the first problem has the same pore pressure distribution as in the sample of the testing device.

Dewatering Below a Clay Layer

We consider the problem of using wells for dewatering below a clay layer as the final example of one-dimensional consolidation. The dewatering causes reduction of pressure below the clay layer. The author was consulted by an engineer from a dewatering company, Pennings [2018], who expressed concern for possible immediate settlement resulting from the pressure reduction. The concern was that pressure can react instantly to a change in load, causing settlements. This idea does not apply in this case; the total stress does not change as there is no change in the loading of the clay layer.

It follows from Terzaghi's definition of effective stress and (10.17) on page 163, that

$$\overset{c}{\sigma}_{zz} = 0 = \overset{c}{\sigma}_{zz}{}' + \overset{c}{p} \rightarrow \overset{c}{p} = -\overset{c}{\sigma}_{zz}{}' = \frac{e_{zz}}{m_v} \tag{10.54}$$

Since the volume strain can change only upon flow, which takes time, the change in pore pressure is initially zero,

$$\overset{c}{p}(z,0) = 0 \quad \rightarrow \quad P(z,0) = 0, \tag{10.55}$$

The boundary conditions are that the pressure change at the top of the layer, $z = 0$, is zero, and at the bottom, $z = -H$, is equal to minus the amount that the pore pressure is lowered, $-\overset{0}{p}$. We transform the boundary condition at $z = -H$, to $\tilde{P} = -\overset{0}{p}/s$,

$$\begin{aligned} P(0,t) &= 0 \\ P(-H,t) &= -\frac{\overset{0}{p}}{s} \end{aligned} \tag{10.56}$$

We choose $\beta = 0$ in (10.37) and apply the first boundary condition, which gives

$$B = -A. \tag{10.57}$$

so that the expression for $\tilde{P}$ becomes

$$\tilde{P} = 2A\sinh(\alpha z) \tag{10.58}$$

We use this in applying the second boundary condition,

$$2A\sinh(-\alpha H) = -\frac{\overset{0}{p}}{s} \rightarrow A = \frac{1}{2}\frac{\overset{0}{p}}{s}\sinh(\alpha H), \tag{10.59}$$

and the final expression for $\tilde{P}$ becomes

$$\tilde{P} = \frac{\overset{0}{p}}{s}\frac{\sinh(\alpha z)}{\sinh(\alpha H)} = \frac{\overset{0}{p}}{s}\frac{\sinh(\sqrt{s/c_v}z)}{\sinh(\sqrt{s/c_v}H)}. \tag{10.60}$$

Since there is no applied load, $\overset{c}{\sigma}_{zz} = 0$, $P = \overset{c}{p}$, and we obtain the inverse transformation of (10.60) from a table,

$$\overset{c}{p} = \overset{0}{p}\left\{\frac{z}{H} + \frac{2}{\pi}\sum_{n=1}^{\infty}\frac{(-1)^n}{n}\mathrm{e}^{-n^2\pi^2\tau}\sin(n\pi Z)\right\}. \tag{10.61}$$

where

$$\tau = \frac{tc_v}{H^2} \qquad Z = \frac{z}{H}. \tag{10.62}$$

We see from (10.61) that $\overset{c}{p} = 0$ at $z = 0$, and $\overset{c}{p} = -\overset{0}{p}$ at $z = -H$.

Curves of $\overset{c}{p}/\overset{0}{p}$ versus Z for constant values of time are shown in Figure 10.5. Pressure reduction initially occurs mostly in the lower part of the clay layer; the final pore pressure change is linear, which corresponds to downward flow through the clay layer.

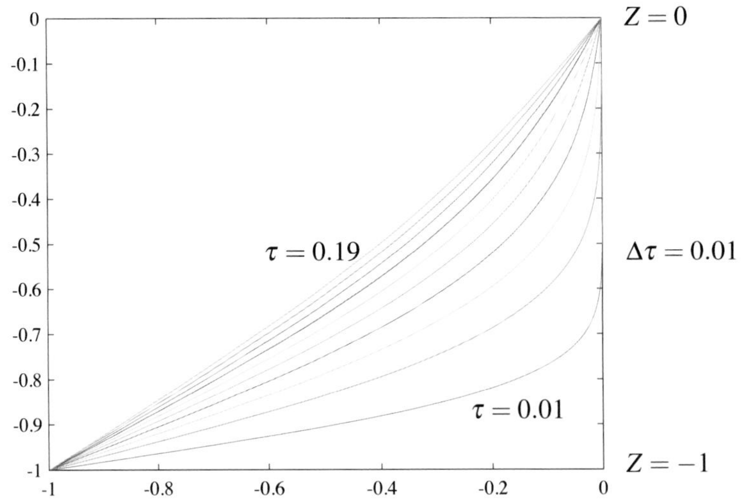

Figure 10.5: Curves of $\overset{c}{p}/\overset{0}{p}$ versus Z for constant values of τ. The interval between curves is $\Delta\tau = 0.01$; the starting time is $\tau = 0.01$ and the final time is $\tau = 0.19$. The final distribution of $\overset{c}{p}/\overset{0}{p}$ is a straight line

10.6.2 Transient Groundwater Flow

We apply the Laplace transform approach to two cases of groundwater flow, both concerning a well that pumps a discharge Q from an aquifer.

A Transient Well Switched on at Time Zero

We consider flow of groundwater caused by a fully penetrating well that is switched on at time $t = 0$, and is placed in a horizontal porous layer of constant thickness H, bounded by two impermeable layers, see Figure 10.6. The problem is equivalent to that of a heat source in a vertical rod placed in an infinite plate of conductive material. The governing equation is (10.11), which we transform to

$$\nabla^2\tilde{\Phi} = \frac{S_s}{k}\left[s\tilde{\Phi} - \Phi(r,0)\right], \tag{10.63}$$

where r is a radial coordinate with its origin at the well center. We determine the change in potential from its original value, which we set at zero, i.e., $\Phi(r,0) = 0$, so that

$$\nabla^2\tilde{\Phi} = \alpha s\tilde{\Phi}, \tag{10.64}$$

where

$$\alpha = \frac{S_s}{k} \tag{10.65}$$

Equation (10.64) is the modified Bessel equation.

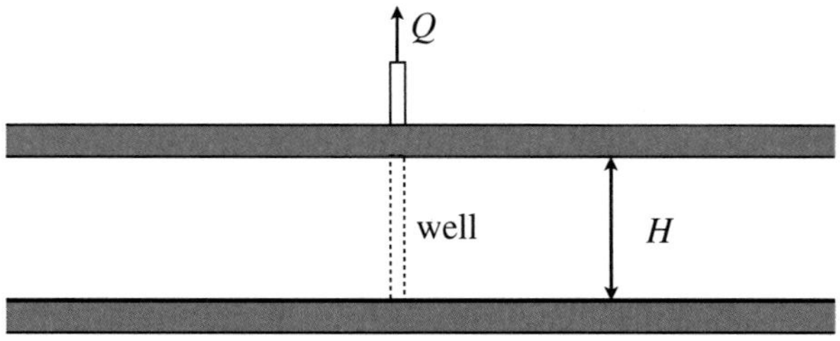

Figure 10.6: A well in a confined aquifer switched on at time $t = 0$

We let the discharge of the well increase from zero to Q instantly at time $t = 0$. The Laplace transform is ideally suited to model such an abrupt increase, emulating the nearly immediate effect of switching on the pump. The solutions obtained with the Laplace transform are a function of time that is equal to zero at time 0^-, i.e., just before the moment the effect takes place. We apply the Heaviside unit step function to the discharge of the well, which transforms as:

$$\tilde{Q} = \frac{Q}{s} \tag{10.66}$$

We restrict the solution to a well of radius r_w that is so small that $\sqrt{\alpha s} r_w << 1$. In that case, an accurate expression for the discharge potential of the transformed equation is, see Strack [2017],

$$\tilde{\Phi} = -\frac{Q}{2\pi s}\mathrm{K}_0(\sqrt{\alpha s} r), \tag{10.67}$$

where K_0 is the modified Bessel function of the second kind and order zero. The inverse transformation exists, and is given in Carslaw and Jaeger [1959],

$$\Phi = -\frac{Q}{4\pi}\int_u^\infty \frac{\mathrm{e}^{-u}}{u} du = -\frac{Q}{4\pi}E_1(u), \tag{10.68}$$

where $E_1(u)$ is the exponential integral, and

$$u(r,t) = \frac{\alpha r^2}{4t} = \frac{S_s r^2}{4kt}. \tag{10.69}$$

10.6.3 Two-Dimensional Instantaneous Sink

The potential for an instantaneous two-dimensional sink is useful to obtain solutions for a sink with a discharge that varies over time, by means of integration. In this case, the discharge is represented by the Dirac delta function, i.e.,

$$Q_0 = \delta(t,0)Q, \tag{10.70}$$

where Q_0 is the discharge pumped over the infinitely short period of time. The transformed discharge is

$$\tilde{Q}_0 = Q_0 \tag{10.71}$$

where we used the transformed Dirac delta function which is 1, see (10.28). The transformed discharge potential becomes

$$\tilde{\Phi} = -\frac{Q}{2\pi}\mathrm{K}_0(\sqrt{\alpha s}r) \tag{10.72}$$

The transform of this function is also given in Carslaw and Jaeger [1959], and is:

$$\Phi = kH\phi = -\frac{Q}{4\pi t}\mathrm{e}^{-\alpha r^2/(4t)} = -\frac{Q}{4\pi t}\mathrm{e}^{-S_s r^2/(4kt)}, \tag{10.73}$$

so that

$$\phi = -\frac{Q}{4\pi kHt}\mathrm{e}^{-S_s r^2/(4kt)}. \tag{10.74}$$

Instantaneous Heat Source

We apply this solution to the flow of heat, with an instantaneous source of an amount Q. ϕ must be replaced by the temperature, T, the thickness of the aquifer, H, becomes 1, k is the thermal conductivity, and the sign of Q must be changed to $-$, thus

$$T = \frac{Q}{4\pi kt}\mathrm{e}^{-r^2/(4\kappa t)} \tag{10.75}$$

This equation corresponds to the one given on page 258 of Carslaw and Jaeger [1959].

Problem 10.1

Consider the problem of diffusion in a fluid with a contaminant of concentration c [M/L^3], injected at time $t = 0$. The concentration is due to an instantaneous source at time $t = t_0$. The problem is one-dimensional; spreading occurs only in the x-direction. The concentration is zero everywhere initially, except at $x = x_0$

Use the solution presented above for an instantaneous heat source, and answer the following questions:

1. Modify the parameters in (10.75) so that it applies to the diffusion equation, and demonstrate that the result satisfies that equation.
2. Modify the solution for the instantaneous two-dimensional source to a one-dimensional one and show that the solution satisfies the differential equation, letting Q represent the injection of mass per unit length normal to the x-direction.
3. Sketch the curves $c = 0.75m/\sqrt{2\pi}$, $c = 0.5m/\sqrt{2\pi}$, and $c = 0.25m/\sqrt{2\pi}$ in the x,τ plane. Do this by setting $c =$ constant in your solution, solving for x in terms of τ, and computing points of the curve (at least 5 points per curve).

Problem 10.2

Derive the potential for an instantaneous sink by subtracting the potential for a sink, (10.68), that begins at time $t + \Delta t$ from a source with the same discharge, but beginning at time t. Take the limit for the discharge Q times the time interval, Δt to be one, to obtain the potential for an instantaneous source of unit discharge injected at time t. Compare your result of that given in the text for an instantaneous source.

10.7 Separation of Variables

Another method for solving the heat equation is the method of *separation of variables*. We restrict the approach to problems where the solution can be written in terms of the product of two functions, one of the spatial variable(s), and the other of time. We illustrate this approach to another problem of groundwater flow in an aquifer, subjected to tidal fluctuation along a river bank. In contrast to the Laplace transform approach, the boundary condition, usually in terms of a periodic function, applies over all times, rather than beginning at some given time.

10.7.1 Response to a Sinusoidal Tidal Fluctuation

We consider the case of transient groundwater flow in a confined aquifer induced by a fluctuating river stage. The water table at the boundary $(x = 0)$ of the unconfined aquifer has a sinusoidal time variation. The boundary condition along the river bank is that the head in the aquifer is equal to the following expression along the river bank, $x = 0$,

$$\phi(0,t) = \hat{\phi} + \Delta h \sin(\omega t), \tag{10.76}$$

where $\hat{\phi}$ is the average head in the aquifer, and where Δh and $\omega/(2\pi)$ are the amplitude and the frequency of the waves, respectively. The second boundary condition is that the head at infinity is not affected by the river,

$$\lim_{x\to\infty}[h(x,t)] = \hat{h}. \tag{10.77}$$

The latter boundary condition implies that $\hat{h}$ is the average value of h. The governing differential equation for this case of flow is the same as that considered for the transient well, with x replacing r,

$$\frac{\partial^2\Phi}{\partial x^2} = \frac{S_s}{k}\frac{\partial\Phi}{\partial t}, \tag{10.78}$$

where

$$\Phi = kH\phi \tag{10.79}$$

We expect the solution to have a period $2\pi/\omega$, and pose a solution of the form

$$\Phi(x,t) = \Re\{F(x)e^{\mathrm{i}\omega t}\}. \tag{10.80}$$

We assume that the solution can be written as the product of a function of x only, $F(x)$, and a function of t only, $e^{\mathrm{i}\omega t}$, which has a period of $2\pi/\omega$. The functions $e^{\mathrm{i}\omega t}$ and $F(x)$ are both complex. Since $\Phi(x,t)$ is a real function, the solution is the real part of $F(x)e^{\mathrm{i}\omega t}$.

We require that the complex function,

$$f(x,t) = F(x)e^{\mathrm{i}\omega t}, \tag{10.81}$$

fulfills the differential equation (10.78). We substitute the expression for $f(x,t)$ in the differential equation,

$$e^{\mathrm{i}\omega t}\frac{d^2F}{dx^2} = \frac{s_p}{k\hat{h}}(\mathrm{i}\omega)Fe^{\mathrm{i}\omega t}, \tag{10.82}$$

and divide by $e^{i\omega t}$,

$$\frac{d^2F}{dx^2} = i\frac{s_p}{k\hat{h}}\omega F. \tag{10.83}$$

We simplify this equation by introducing a factor λ,

$$\frac{d^2F}{dx^2} - \frac{F}{\lambda^2} = 0, \tag{10.84}$$

where λ is a complex constant,

$$\frac{1}{\lambda} = \sqrt{i\frac{s_p\omega}{k\bar{h}}} = \sqrt{\frac{s_p\omega}{k\bar{h}}e^{i\pi/2}} = \sqrt{\frac{s_p\omega}{k\bar{h}}}e^{i\pi/4} = \sqrt{\frac{s_p\omega}{k\bar{h}}}\left[\frac{1}{\sqrt{2}} + \frac{i}{\sqrt{2}}\right]. \tag{10.85}$$

The differential equation (10.84) for $F(x)$ has the general solution:

$$F(x) = c_1 e^{x/\lambda} + c_2 e^{-x/\lambda}. \tag{10.86}$$

Both λ and F are complex, as are the coefficients c_1 and c_2. We introduce a new parameter μ:

$$\frac{1}{\lambda} = \sqrt{\frac{s_p\omega}{2k\bar{h}}}(1+i) = \mu(1+i), \tag{10.87}$$

where

$$\mu = \sqrt{\frac{s_p\omega}{2k\bar{h}}}. \tag{10.88}$$

We substitute (10.87) for $1/\lambda$ in (10.86) ,

$$F = c_1 e^{\mu(1+i)x} + c_2 e^{-\mu(1+i)x}, \tag{10.89}$$

and expression (10.80) for Φ becomes

$$\Phi = \Re\{c_1 e^{\mu(1+i)x} e^{i\omega t}\} + \Re\{c_2 e^{-\mu(1+i)x} e^{i\omega t}\}, \tag{10.90}$$

or

$$\Phi = \Re\{c_1 e^{\mu x} e^{i(\omega t+\mu x)}\} + \Re\{c_2 e^{-\mu x} e^{i(\omega t-\mu x)}\}. \tag{10.91}$$

We determine the real part,

$$\begin{aligned}\Phi =& e^{\mu x}[\Re(c_1)\cos(\omega t+\mu x) - \Im(c_1)\sin(\omega t+\mu x)]\\ &+ e^{-\mu x}[\Re(c_2)\cos(\omega t-\mu x) - \Im(c_2)\sin(\omega t-\mu x).\end{aligned} \tag{10.92}$$

This equation represents the general solution for the class of problems defined by a periodic boundary condition applied at some given value for x in a semi-infinite aquifer. The next step is to determine the constants in the solution. The boundary condition is at $x = 0$ and the aquifer is defined by $0 \le x < \infty$; the boundary condition is:

$$\Phi(0,t) = \Delta\Phi\sin(\omega t). \tag{10.93}$$

where

$$\Delta\Phi = kH\Delta\phi \tag{10.94}$$

We observe from (10.92) that (10.77) can be satisfied only if c_1 is zero,

$$c_1 = 0. \tag{10.95}$$

Application of boundary condition (10.93) to (10.92) yields, with $c_1 = 0$,

$$\Delta\Phi \sin(\omega t) = \Re(c_2)\cos(\omega t) - \Im(c_2)\sin(\omega t), \tag{10.96}$$

so that

$$\Im(c_2) = -\Delta\Phi \qquad \Re(c_2) = 0, \tag{10.97}$$

and

$$c_2 = -\mathrm{i}\Delta\Phi. \tag{10.98}$$

We substitute zero for c_1 and $-\mathrm{i}\Delta\Phi$ for c_2 in expression (10.92) for Φ,

$$\Phi = \Delta\Phi e^{-\mu x}\sin(\omega t - \mu x) \tag{10.99}$$

The assumption that Φ can be represented by (10.80) appears to be correct; both the differential equation and the boundary conditions are fulfilled.

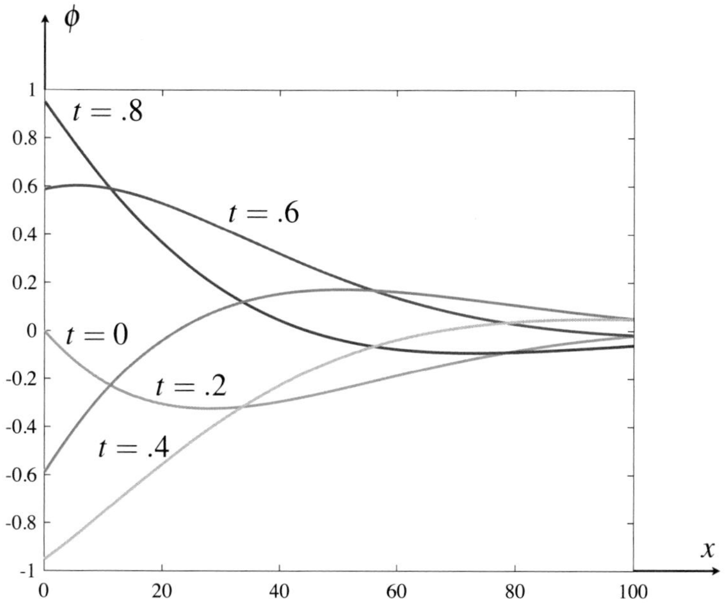

Figure 10.7: Curves of the hydraulic head versus x for different values of time. The data are: $s_p = 0.05$, $k = 10$ m/d, $h = 20$ m, $H = 10$ m, $\Delta h = 1$ m, $\omega = 2 * \pi$ 1/d. Distance is in meters, and time in days

10.7.2 Initial Condition

The solution obtained using separation of variables is fundamentally different from that obtained using the Laplace transform. In contrast to the Laplace transform approach, the initial condition cannot be specified; the solution continues throughout time indefinitely and has neither a beginning, nor an end. Such solutions may be useful if an initial condition is not essential, but it is important to be aware of this limitation of the method.

Chapter 11

The Hyperbolic Case: Two Real Characteristics

As an illustration of a set of two linear hyperbolic partial differential equations, we consider first longitudinal vibration in a bar, and second transverse vibration in a string. We begin by deriving the equations that govern these two cases of vibration, which are similar in form. We examined the coefficients in these equations in Chapter 7 and established that the system is hyperbolic. We write the differential equations in the characteristic directions, and both determine and examine the solution.

11.1 Longitudinal Vibration in a Bar

We consider a steel beam of cross-sectional area A and Young's modulus E, shown in Figure 11.1. The increase in length Δw of an incremental section of initial length Δx is:

$$\Delta w = w(x + \tfrac{1}{2}\Delta x, t) - w(x - \tfrac{1}{2}\Delta x, t), \tag{11.1}$$

where w is the displacement in the x-direction. The strain,

$$\varepsilon = \lim_{\Delta x \to 0} \frac{\Delta w}{\Delta x} = \frac{\partial w}{\partial x}, \tag{11.2}$$

is related to the longitudinal force F in the beam by

$$\frac{F}{A} = \varepsilon E = E \frac{\partial w}{\partial x}. \tag{11.3}$$

We apply Newton's second law to the mass $\rho A \Delta x$,

$$F(x + \tfrac{1}{2}\Delta x, t) - F(x - \tfrac{1}{2}\Delta x, t) = \rho A \Delta x \frac{\partial^2 w}{\partial t^2}, \tag{11.4}$$

and pass to the limit for $\Delta x \to 0$,

$$\frac{\partial F}{\partial x} = \rho A \frac{\partial^2 w}{\partial t^2}, \tag{11.5}$$

O. D. L. Strack, *Applications of Vector Analysis and Complex Variables in Engineering*, https://doi.org/10.1007/978-3-030-41168-8_11

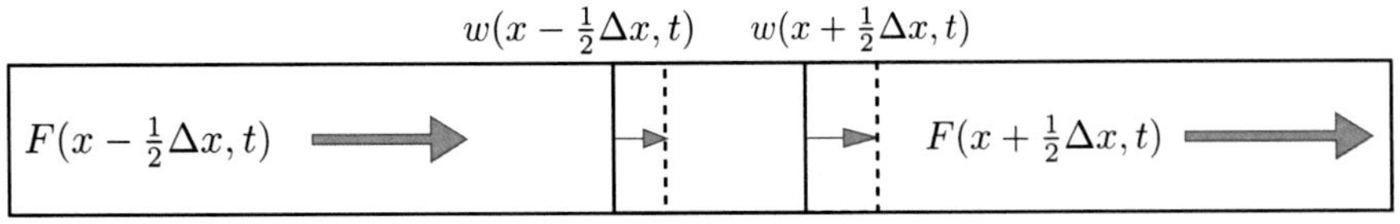

Figure 11.1: Longitudinal vibrations in a bar

combine this with (11.3)

$$AE\frac{\partial^2 w}{\partial x^2} = \rho A\frac{\partial^2 w}{\partial t^2}, \tag{11.6}$$

and divide by AE,

$$\frac{\partial^2 w}{\partial x^2} - \frac{\rho}{E}\frac{\partial^2 w}{\partial t^2} = 0. \tag{11.7}$$

11.2 Transverse Vibration in a String

We consider a vibrating string as a second application. A section of string is shown in Figure 11.2. The string has a mass m per unit length of string [M/L], and is subject to an axial force S which is assumed to have a constant horizontal component. An elementary section of string of length Δl is inclined at an angle α to the x-axis, where α is small; $\alpha \approx \sin\alpha \approx \tan\alpha$ and $\cos\alpha \approx 1$. We neglect the effect of gravity relative to that of S; there is a net upward force of magnitude $F_y = S[\sin\alpha(x+\frac{1}{2}\Delta x,t) - \sin\alpha(x-\frac{1}{2}\Delta x,t)]$. Since α is small, the sines may be replaced by tangents, and $\tan\alpha = \partial y/\partial x$ where y is the elevation of a point of the string. We apply Newton's second law in the vertical direction:

$$\frac{\partial^2 y}{\partial t^2} = \frac{F_y}{m\Delta l} \approx \frac{S}{m}\frac{\tan\alpha(x+\frac{1}{2}\Delta x,t) - \tan\alpha(x-\frac{1}{2}\Delta x,t)}{\Delta x}, \tag{11.8}$$

where $\Delta x = \Delta l\cos\alpha \approx \Delta l$. We pass to the limit for $\Delta x \to 0$,

$$\frac{\partial^2 y}{\partial t^2} = \frac{S}{m}\frac{\partial}{\partial x}\tan\alpha = \frac{S}{m}\frac{\partial^2 y}{\partial x^2}, \tag{11.9}$$

or

$$\frac{\partial^2 y}{\partial x^2} - \frac{m}{S}\frac{\partial^2 y}{\partial t^2} = 0. \tag{11.10}$$

The equation for the vibrating string is identical in form to that of the longitudinal vibrations in a bar.

11.3 The Differential Equation Along the Characteristics

We are now in a position to examine the type of the partial differential equations. We refer to the string during our analysis, because transverse vibrations are easier to imagine than longitudinal ones. Both (11.7) and (11.10) have the form:

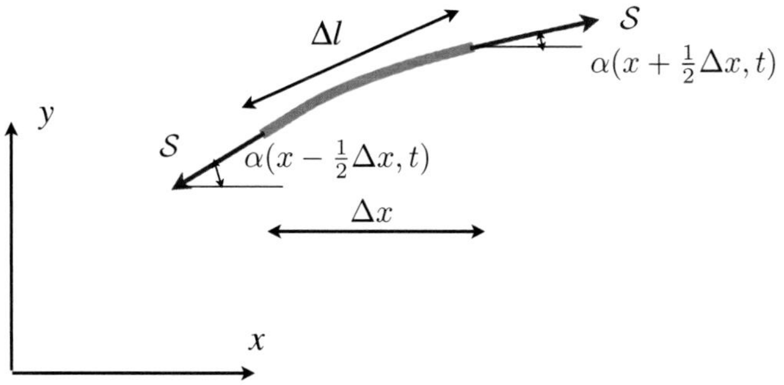

Figure 11.2: Transverse vibration in a string

$$\frac{\partial^2 y}{\partial x^2} - \frac{1}{c^2}\frac{\partial^2 y}{\partial t^2} = 0, \tag{11.11}$$

which is the one-dimensional wave equation. We simplify the differential equation further with the substitution

$$\tau = ct \qquad [\mathrm{L}^2] \tag{11.12}$$

and (11.11) becomes:

$$\frac{\partial^2 y}{\partial x^2} - \frac{\partial^2 y}{\partial \tau^2} = 0. \tag{11.13}$$

In order to bring this differential equation in the form of two first-order partial differential equations, we make the substitutions:

$$u = \frac{\partial y}{\partial x} \qquad v = \frac{\partial y}{\partial \tau} \tag{11.14}$$

so that (11.13) becomes:

$$\frac{\partial u}{\partial x} - \frac{\partial v}{\partial \tau} = 0. \tag{11.15}$$

The function $y = y(x, \tau)$ is single-valued, so that the order of differentiation is immaterial; (11.14) implies that

$$\frac{\partial u}{\partial \tau} - \frac{\partial v}{\partial x} = 0 \tag{11.16}$$

We analyzed the wave equation in section 7.2 and obtained the following characteristic directions in (7.35) and (7.36) on page 81:

$$\tan \underset{1}{\alpha} = -1 \rightarrow \underset{1}{\alpha} = -\frac{\pi}{4} \qquad s_1 - \text{characteristic} \tag{11.17}$$

and

$$\tan \underset{2}{\alpha} = 1 \rightarrow \underset{2}{\alpha} = \frac{\pi}{4} \qquad s_2 - \text{characteristic} \tag{11.18}$$

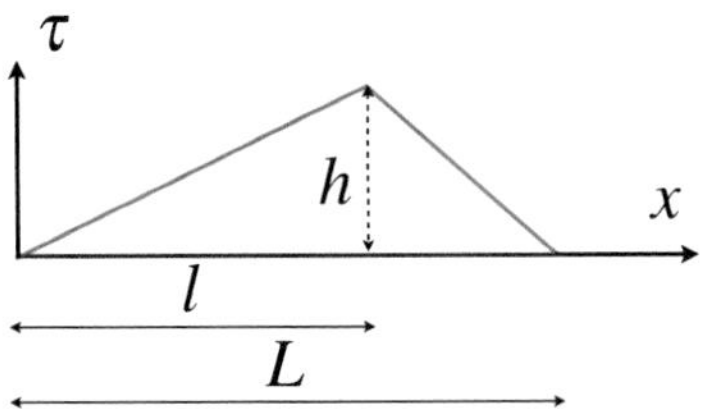

Figure 11.3: Initial shape of the string

and the equation along the characteristics are given by (7.42) and (7.43):

$$u+v=\underset{1}{c} \qquad \text{(along } s_1\text{)}, \tag{11.19}$$

and

$$u-v=\underset{2}{c} \qquad \text{(along } s_2\text{)}, \tag{11.20}$$

where $\underset{1}{c}$ and $\underset{2}{c}$ are constants. The sum of u and v is constant along the s_1 characteristics, whereas their difference is constant along the s_2-characteristics.

11.4 Initial Value Problem With Discontinuous Shape

We consider the following initial value problem. At time $\tau = 0$, the string is in the position shown in Figure 11.3:

$$\tau = 0 \qquad 0 \le x \le l \qquad u = \frac{\partial y}{\partial x} = \frac{h}{l} = m \tag{11.21}$$

$$\tau = 0 \qquad l \le x \le L \qquad u = \frac{\partial y}{\partial x} = \frac{-h}{L-l} = -n. \tag{11.22}$$

The string has no initial velocity at $\tau = 0$, and remains fixed at all times at the end points $x = 0$ and $x = L$:

$$0 \le \tau < \infty \qquad x = 0 \qquad v = \frac{\partial y}{\partial \tau} = 0 \tag{11.23}$$

$$0 \le \tau < \infty \qquad x = L \qquad v = \frac{\partial y}{\partial \tau} = 0 \tag{11.24}$$

$$\tau = 0 \qquad 0 \le x \le L \qquad v = \frac{\partial y}{\partial \tau} = 0. \tag{11.25}$$

These boundary conditions, along with certain characteristics are indicated in the x, τ diagram of Figure 11.4, which is a square of side L.

The s_1- and s_2- characteristics are shown that start either at a corner point of the square or at the discontinuity in u at $x = l$. The equations (11.19) and (11.20) are shown in the figure as they apply along the s_1- and s_2- characteristics.

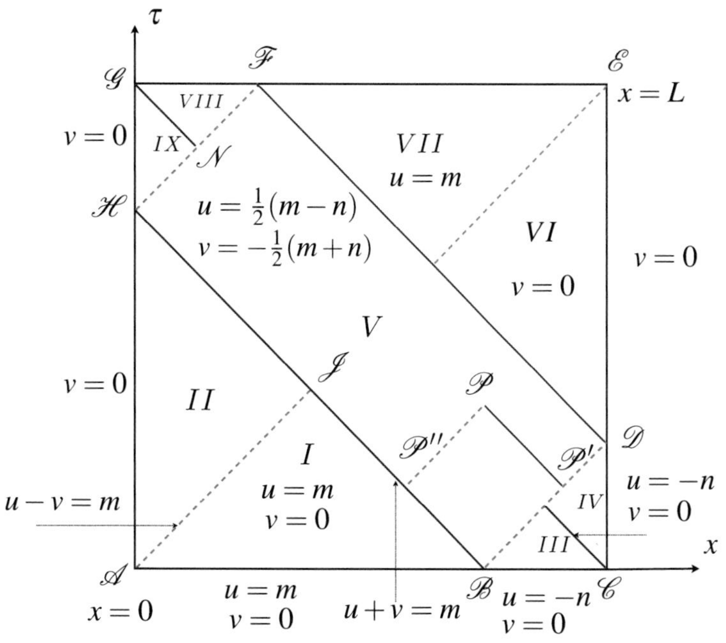

Figure 11.4: Characteristics and initial values for the vibrating string

We determine the solution in steps. The procedure is governed by the boundary conditions and the pattern of characteristics. We observe that the solution in domain I is fully determined by the boundary values along $\mathscr{A}\mathscr{B}$. We find the solution at any point inside I by finding the two characteristics that intersect at that point, tracing them back to the boundaries, and using (11.19) and (11.20). We apply this procedure to point $\mathscr{J}$, which is the point of intersection of the s_1 characteristic through $\mathscr{B}$ and the s_2 characteristic through $\mathscr{A}$; we have two conditions at point $\mathscr{J}$:

$$\text{Point } \mathscr{J}: \; u+v=m \qquad u-v=m, \tag{11.26}$$

so that

$$\text{Point } \mathscr{J}: \; u=m \qquad v=0. \tag{11.27}$$

We apply the same procedure to all points inside domain I:

$$\text{Domain } I: \; u=m \qquad v=0. \tag{11.28}$$

We have along $\mathscr{A}\mathscr{J}$ that $u-v=m-0=\text{constant}$. The boundary-value problem in domain I is called the *Cauchy problem*; in this problem the values of both functions are known along a boundary section that does not coincide with a characteristic. In the Cauchy problem, the solution is determined by the boundary conditions only in the region enclosed by the boundary and two boundary characteristics ($\mathscr{A}\mathscr{J}$ and $\mathscr{B}\mathscr{J}$ in this case). In domain II, the values of u and v are both known along $\mathscr{A}\mathscr{J}$: $u=m$

and $v=0$. The other boundary condition for domain II is that $v=0$ along $\mathscr{A}\mathscr{H}$. The boundary value problem in domains II is known as the *mixed boundary value problem*: one of the functions is given along a non-characteristic ($\mathscr{A}\mathscr{H}$) and both are given along a characteristic ($\mathscr{A}\mathscr{J}$). We find the solution by intersecting the s_1-characteristics ($u+v=$constant) starting at points of $\mathscr{A}\mathscr{J}$ with $\mathscr{A}\mathscr{H}$. Since $u=m$, $v=0$ along $\mathscr{A}\mathscr{J}$, $u+v=m$ along the s_1-characteristics, we have at the intersection with $\mathscr{A}\mathscr{H}$ that $u+v=u=m$. Hence:

$$\text{Domain } II: \; u=m \quad v=0. \tag{11.29}$$

In a similar fashion, we find for domains III and IV:

$$\text{Domain } III: \; u=-n \quad v=0, \tag{11.30}$$

and

$$\text{Domain } IV: \; u=-n \quad v=0. \tag{11.31}$$

In domain V the boundary values of both u and v are known along the bounding characteristics $\mathscr{B}\mathscr{D}$ and $\mathscr{B}\mathscr{H}$. Such a boundary value problem is called the *characteristic boundary-value problem* or *Riemann problem.* We must remember in dealing with characteristics that emenate from a point of discontinuty ($\mathscr{B}$ in this case) that the values of both u and v jump. These jumps are such that $u+v$ is constant along s_1 *across* $\mathscr{B}\mathscr{D}$ and $u-v$ is constant along s_2 *along* $\mathscr{B}\mathscr{D}$. Hence, to find u and v at a point $\mathscr{P}$, we must intersect the s_1 characteristic $\mathscr{P}'\mathscr{P}$ (where $u+v=-n$) with the s_2 characteristic $\mathscr{P}''\mathscr{P}$ (where $u-v=m$). At point $\mathscr{P}$, therefore we have both that

$$u-v=m, \tag{11.32}$$

and

$$u+v=-n. \tag{11.33}$$

Hence,

$$u=\ \tfrac{1}{2}(m-n), \tag{11.34}$$

and

$$v=-\tfrac{1}{2}(m+n). \tag{11.35}$$

Since this is true for any point $\mathscr{P}$ in domain V, we have

$$\text{Domain } V: \; u=\tfrac{1}{2}(m-n) \quad v=-\tfrac{1}{2}(m+n). \tag{11.36}$$

We find the solution in domains VI through IX using the known values of u and v,

$$\begin{aligned} &\text{Domain } VI: \quad u=m \quad v=0\\ &\text{Domain } VII: \quad u=m \quad v=0\\ &\text{Domain } VIII: \quad u=-n \quad v=0\\ &\text{Domain } IX: \quad u=-n \quad v=0 \end{aligned} \tag{11.37}$$

It appears that $u=\partial y/\partial x$ is a constant and $v=\partial y/\partial t$ is zero everywhere except in domain V, where the slope $\partial y/\partial x$ is $\frac{1}{2}(m-n)$ and points of the string move downward

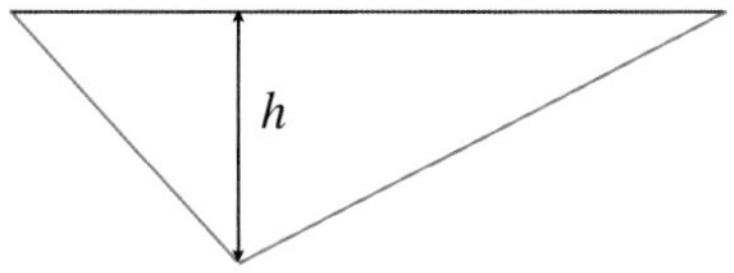

Figure 11.5: The string at time $\tau = ct = L$

with a velocity $\frac{1}{2}(m+n)$. At time $\tau = ct = L$, $\mathscr{G}\mathscr{F}\mathscr{E}$ in Figure 11.4, the string has attained the shape shown in Figure 11.5 and the value of y at point $\mathscr{F}$ is $-h$. We continue he solution beyond $\tau = L$ and find that points of the string start moving up again.

Problem 11.1

Consider the case of a plucked string; the initial shape of the string is a horizontal line, except at some point between the fixed points of the string, where a single point of the string is held above the rest of the string. This case is the limiting case of the slope of the string in Figure 11.4 being zero up to a point at a distance l from the left fixed side, where the slope becomes infinite. The elevation of the string to the right of the point where it is plucked is zero. The problem is illustrated in Figure 11.6.

Question:
Determine the shape of the string as a function of time and place in the same manner as was done for the vibrating string, creating a diagram similar to that of Figure 11.4.

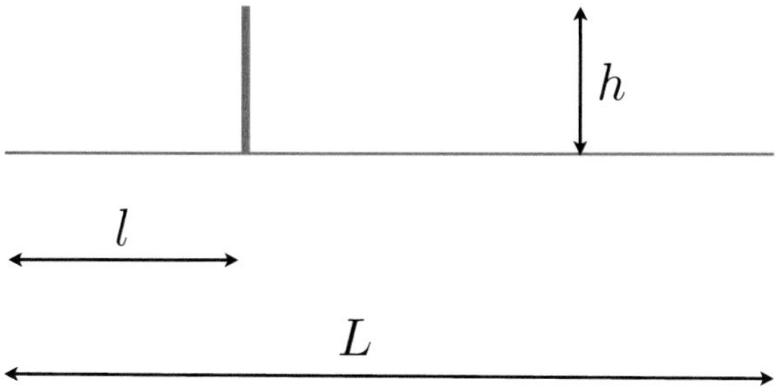

Figure 11.6: A plucked string

11.5 Initial Value Problem With Smooth Shape

We consider the case of vibration of a string or rod, but now with an initial shape that is continuous. The initial conditions for $u = \partial y/\partial \tau$ and $v = \partial y/\partial \tau$ are:

$$
\begin{aligned}
u &= \frac{\partial y}{\partial x} = h \sin\left(\frac{\pi}{L}x\right) \qquad && \tau = 0 \\
v &= \frac{\partial y}{\partial \tau} = 0 && \tau = 0
\end{aligned}
\tag{11.38}
$$

The boundary conditions are that the ends at $x = 0$ and $x = L$ are fixed, so that $v = 0$,

$$
\begin{aligned}
v &= 0 \qquad && x = 0 \\
v &= 0 && x = L
\end{aligned}
\tag{11.39}
$$

We solve the problem first using the method of characteristics, and then by applying the Laplace transform.

11.5.1 Solution Using Characteristics

We write the solution in terms of the equations along the characteristics. We introduce the (ξ, η)-Cartesian coordinates along the characteristics derived in Chapter 7, illustrated in Figure 11.7:

$$
\begin{aligned}
\xi &= \tau - x \\
\eta &= \tau + x
\end{aligned}
\tag{11.40}
$$

The equations along the characteristics are given in equations (7.42) and (7.43) on page 83,

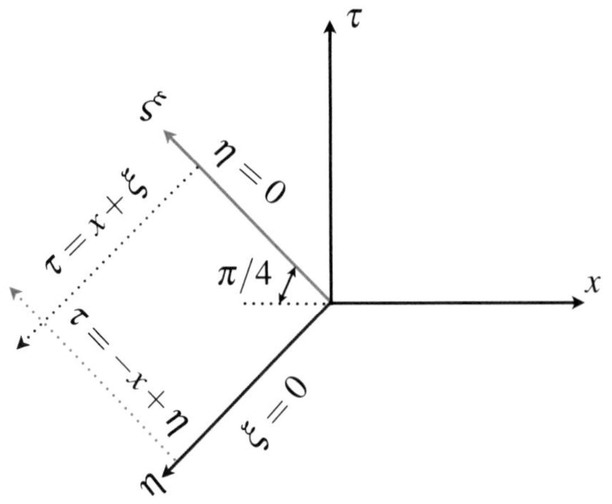

Figure 11.7: The characteristic directions and the coordinates along them

$$\begin{aligned} u+v &= f(\eta) \quad \text{along} \quad \xi; \quad \eta = \text{constant} \\ u-v &= g(\xi) \quad \text{along} \quad \eta; \quad \xi = \text{constant}. \end{aligned} \tag{11.41}$$

By definition, η is constant along the ξ-characteristics, and ξ is constant along the η-characteristics, so that $u+v$ can only be a function of η, and $u-v$ can only be a function of ξ. We solve the expressions for $u+v$ and $u-v$ given in (11.41) for u and v:

$$\begin{aligned} u &= \tfrac{1}{2}[g(\xi)+f(\eta)] = \tfrac{1}{2}[g(\tau-x)+f(\tau+x)] \\ v &= \tfrac{1}{2}[-g(\xi)+f(\eta)] = \tfrac{1}{2}[-g(\tau-x)+f(\tau+x)]. \end{aligned} \tag{11.42}$$

Application of the Initial and Boundary Conditions

We first apply the condition that $v=0$ initially:

$$-g(-x)+f(x)] = 0, \tag{11.43}$$

then apply the condition that $v=0$ for $x=0$,

$$-g(\tau)+f(\tau) = 0. \tag{11.44}$$

The combination of the latter two equations gives:

$$\begin{aligned} f(x) &= g(x) \\ f(x) &= f(-x), \end{aligned} \tag{11.45}$$

so that the expressions for u and v are:

$$\begin{aligned} u &= \tfrac{1}{2}[f(x-\tau)+f(x+\tau)] \\ v &= \tfrac{1}{2}[-f(x-\tau)+f(x+\tau)]. \end{aligned} \tag{11.46}$$

We apply the initial condition for u, (11.38),

$$u(x,0) = \frac{\partial y}{\partial x} = \frac{\pi}{L} h \cos\left(\frac{\pi x}{L}\right) = f(x), \tag{11.47}$$

so that

$$\begin{aligned} u &= \frac{\partial y}{\partial x} = \frac{\pi h}{2L}\left[\cos\left(\frac{\pi(x+\tau)}{L}\right) + \cos\left(\frac{\pi(x-\tau)}{L}\right)\right] \\ v &= \frac{\partial y}{\partial \tau} = \frac{\pi h}{2L}\left[\cos\left(\frac{\pi(x+\tau)}{L}\right) - \cos\left(\frac{\pi(x-\tau)}{L}\right)\right]. \end{aligned} \tag{11.48}$$

We split the cosines into a sum of cosines and sines of either only x or only τ:

$$\begin{aligned} \frac{\partial y}{\partial x} &= \frac{\pi h}{2L}\left[\cos\frac{\pi x}{L}\cos\frac{\pi\tau}{L} - \sin\frac{\pi x}{L}\sin\frac{\pi\tau}{L} + \cos\frac{\pi x}{L}\cos\frac{\pi\tau}{L} + \sin\frac{\pi x}{L}\sin\frac{\pi\tau}{L}\right] \\ \frac{\partial y}{\partial \tau} &= \frac{\pi h}{2L}\left[\cos\frac{\pi x}{L}\cos\frac{\pi\tau}{L} - \sin\frac{\pi x}{L}\sin\frac{\pi\tau}{L} - \cos\frac{\pi x}{L}\cos\frac{\pi\tau}{L} - \sin\frac{\pi x}{L}\sin\frac{\pi\tau}{L}\right] \end{aligned} \tag{11.49}$$

or

$$\begin{aligned}\frac{\partial y}{\partial x} &= \frac{\pi h}{L}\cos\frac{\pi x}{L}\cos\frac{\pi\tau}{L}\\ \frac{\partial y}{\partial \tau} &= -\frac{\pi h}{L}\sin\frac{\pi x}{L}\sin\frac{\pi\tau}{L}.\end{aligned} \tag{11.50}$$

We obtain the expression for the amplitude of the string, y, by integration of the equations, which gives

$$y = h\sin\frac{\pi x}{L}\cos\frac{\pi\tau}{L}. \tag{11.51}$$

A plot of the string at various times is shown in Figure 11.8.

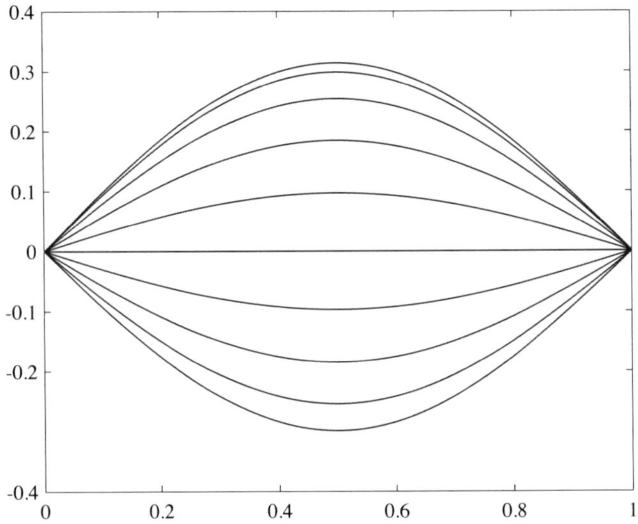

Figure 11.8: Position of the string; the amplitude is in dimensionless form: h/L, and the horizontal axis is x/L. the time interval is $\Delta t/L = 0.1$

11.5.2 Solution Using the Laplace Transform

We illustrate the application of the Laplace transform to a differential equation that involves the second order derivative with respect to time for the case of a moving string with a sinusoidal initial shape. We use the original differential equation, (11.11), for the application of the Laplace transform:

$$\frac{\partial^2 y}{\partial x^2} - \frac{\partial^2 y}{\partial \tau^2} = 0. \tag{11.52}$$

We need the Laplace transform of the second-order time derivative. We have, from the definition of the Laplace transform,

$$\mathscr{L}(f'') = \int_0^\infty f''(\tau)\mathrm{e}^{-s\tau}d\tau = \int_0^\infty \mathrm{e}^{-s\tau}df' = \mathrm{e}^{-s\tau}f'(\tau)|_0^\infty + s\int_0^\infty f'\mathrm{e}^{-s\tau}d\tau, \qquad (11.53)$$

where the prime denotes differentiation with respect to time. We assume that the limit at infinity of the first term is zero,

$$\mathscr{L}(f'') = -f'(0) + s\int_0^\infty \mathrm{e}^{-s\tau}df = -f'(0) + sf\mathrm{e}^{-s\tau}|_0^\infty + s\int_0^\infty f\mathrm{e}^{-s\tau}d\tau, \qquad (11.54)$$

so that, assuming that the upper limit in the second term vanishes at infinity,

$$\mathscr{L}(f'') = \tilde{f}'' = s^2\tilde{f} - sf(0) - f'(0). \qquad (11.55)$$

We obtain the same result by applying the Laplace transform formula twice, first to $f(t)$,

$$\mathscr{L}(f) = s\tilde{f} - f(0) \qquad (11.56)$$

and then to $s\tilde{f} - f(0)$:

$$\mathscr{L}(f) = s^2\tilde{f} - sf(0) - f'(0). \qquad (11.57)$$

We apply this to the wave equation (11.52),

$$\frac{d^2\tilde{y}}{dx^2} = s^2\tilde{y} - sy(x,0) - y'(x,0). \qquad (11.58)$$

We consider a string that is initially at rest, and is released from that position at time $\tau = 0$. The initial position is given by a sine, and the string is fixed at the end points. We express the initial and boundary conditions in terms of y, rather than in terms of its derivatives with respect to time and space:

$$\begin{aligned} &0 \le x \le L \qquad y(x,0) = h\sin\left(\frac{\pi x}{L}\right) \\ &0 \le x \le L \qquad y'(x,0) = 0. \end{aligned} \qquad (11.59)$$

The string remains fixed at all times at the end points $x = 0$ and $x = L$:

$$\begin{aligned} y(0,\tau) &= 0 \\ y(L,\tau) &= 0 \end{aligned} \qquad (11.60)$$

The ordinary differential equation to be solved is (11.58), with $y'(x,0) = 0$, and $y(x,0)$ given by (11.59),

$$\frac{d^2\tilde{y}}{dx^2} - s^2\tilde{y} = -sh\sin\left(\frac{\pi x}{L}\right). \qquad (11.61)$$

We obtain the general solution to this differential equation from standard books on mathematics,

$$y = c_1\mathrm{e}^{sx} + c_2\mathrm{e}^{-sx} - \frac{\mathrm{e}^{sx}}{2s}\int \mathrm{e}^{-sx}sh\sin\left(\frac{\pi x}{L}\right)dx + \frac{\mathrm{e}^{-sx}}{2s}\int \mathrm{e}^{sx}sh\sin\left(\frac{\pi x}{L}\right)dx, \quad (11.62)$$

where c_1 and c_2 are constants. We rearrange,

$$y = c_1 e^{sx} + c_2 e^{-sx} - h\frac{e^{sx}}{2}\int e^{-sx}\sin\left(\frac{\pi x}{L}\right)dx + h\frac{e^{-sx}}{2}\int e^{sx}\sin\left(\frac{\pi x}{L}\right)dx. \quad (11.63)$$

We use the following integral,

$$\int e^{ax}\sin(bx)dx = \frac{e^{ax}}{a^2+b^2}\left[a\sin(bx) - b\cos(bx)\right], \quad (11.64)$$

and obtain the expression for $\tilde{y}$ with (11.62),

$$\begin{aligned} y = c_1 e^{sx} + c_2 e^{-sx} &- h\frac{e^{sx}}{2}\frac{e^{-sx}}{s^2+(\pi/L)^2}\left[-s\sin\frac{\pi x}{L} - \frac{\pi}{L}\cos\frac{\pi x}{L}\right] \\ &+ h\frac{e^{-sx}}{2}\frac{e^{sx}}{s^2+(\pi/L)^2}\left[s\sin\frac{\pi x}{L} - \frac{\pi}{L}\cos\frac{\pi x}{L}\right]. \end{aligned} \quad (11.65)$$

We simplify this,

$$y = c_1 e^{sx} + c_2 e^{-sx} + h\frac{s}{s^2+(\pi/L)^2}\sin\frac{\pi x}{L}, \quad (11.66)$$

and apply the boundary conditions that $\tilde{y}$ is zero for $x = 0$ and $x = L$,

$$\begin{aligned} 0 &= c_1 + c_2 \\ 0 &= c_1 e^{sL} + c_2 e^{-sL}, \end{aligned} \quad (11.67)$$

which gives that

$$c_1 = c_2 = 0. \quad (11.68)$$

We apply the inverse transformation to (11.66), and get the same result as that obtained using the method of characteristics:

$$y = h\sin\frac{\pi x}{L}\cos\frac{\pi\tau}{L} \quad (11.69)$$

This solution meets the condition that y has the prescribed form for $\tau = 0$, as well as the condition that y is zero for $x = 0$ and $x = L$.

It is clear from the solution that a third approach to solve this problem is to apply the method of separation of variables.

Problem 11.2

Demonstrate that (11.69) satisfies the wave equation.

Problem 11.3

Program the solution for the string with the sinusoidal initial shape, and plot the string at various times.

Questions:

1. Determine an expression for the time t_1 that it takes for the string to recover its original shape.
2. Plot the string at times 0, $t_1/4$, $t_1/2$, $3t_1/4$, t_1.

Problem 11.4

Apply the method of separation of variables to solve the problem of a string with sinusoidal initial shape, and compare your solution with the one obtained using the Laplace transform approach.

Chapter 12

Hyperbolic Quasi Linear Partial Differential Equations

12.1 Quasi Linear Partial Differential Equations

We call a differential equation quasi linear if the partial derivatives of the dependent variables are multiplied by functions of the dependent variables. This property affects hyperbolic equations significantly, because the characteristics cannot be determined *a priori*, but depend on the solution. This also implies that discontinuities cannot propagate along the characteristics; the shape of the characteristics themselves would be affected by the jump.

12.2 Granular Soils at Impending Failure

A granular medium, such as sand, can loose stability and fail. Landslides and collapsing dams are examples of such failures. We consider a granular medium that is about to fail and assume that a part of the body has reached such a stress state that the entire body is about to lose equilibrium.

The solution presented here is found in most textbooks on soil mechanics, e.g., Verruijt [2017]. A number of solutions to the partial differential equations for a granular medium at failure are found in Sokolovski [1960].

12.2.1 Conditions for Limit Equilibrium

We consider a granular medium at impending failure; stresses in the body meet the equilibrium conditions. These are, for geological materials with compressive stresses taken positive, (9.132),

$$\partial_i \sigma_{ij} = \beta_j. \tag{12.1}$$

O. D. L. Strack, *Applications of Vector Analysis and Complex Variables in Engineering*, https://doi.org/10.1007/978-3-030-41168-8_12

We set the body force equal to zero; for the application considered here, the body force, gravity, plays a minor role. We expand (12.1) with $\beta_j = 0$,

$$\begin{aligned}\frac{\sigma_{xx}}{\partial x} + \frac{\sigma_{xy}}{\partial y} &= 0\\ \frac{\sigma_{xy}}{\partial x} + \frac{\sigma_{yy}}{\partial y} &= 0.\end{aligned} \tag{12.2}$$

The granular material is in a state of limit equilibrium. This condition implies that half the stress deviator, λ, is related to the isotropic stress, $\overset{0}{\sigma}$, and to two properties of the material: the angle of internal friction, ϕ, and the cohesion c as:

$$\lambda = \overset{0}{\rho} \sin\phi, \tag{12.3}$$

where

$$\overset{0}{\rho} = \overset{0}{\sigma} + c\cot\phi. \tag{12.4}$$

12.3 Differential equations for impending failure of granular media

We express the three components of the stress tensor in terms of λ, the principal direction ψ, and the isotropic stress $\overset{0}{\sigma}$, according to (5.77) and (5.79), on page 63:

$$\begin{aligned}\sigma_{xx} &= \overset{0}{\rho}[1+\sin\phi\cos(2\psi)] - c\cot\phi\\ \sigma_{yy} &= \overset{0}{\rho}[1-\sin\phi\cos(2\psi)] - c\cot\phi\\ \sigma_{xy} &= \overset{0}{\rho}\sin\phi\sin(2\psi).\end{aligned} \tag{12.5}$$

We substitute these expressions for the components of the stress tensor in the equilibrium equations (12.2), and obtain a set of two first-order partial differential equations in terms of the two dependent variables $\overset{0}{\rho}$ and ψ:

$$\begin{aligned}&[1+\sin\phi\cos(2\psi)]\frac{\partial\overset{0}{\rho}}{\partial x} + \sin\phi\sin(2\psi)\frac{\partial\overset{0}{\rho}}{\partial y} - 2\overset{0}{\rho}\sin\phi\sin(2\psi)\frac{\partial\psi}{\partial x}\\ &+2\overset{0}{\rho}\sin\phi\cos(2\psi)\frac{\partial\psi}{\partial y} = 0\end{aligned} \tag{12.6}$$

$$\begin{aligned}&\sin\phi\sin(2\psi)\frac{\partial\overset{0}{\rho}}{\partial x} + [1-\sin\phi\cos(2\psi)]\frac{\partial\overset{0}{\rho}}{\partial y} + 2\overset{0}{\rho}\sin\phi\cos(2\psi)\frac{\partial\psi}{\partial x}\\ &+2\overset{0}{\rho}\sin\phi\sin(2\psi)\frac{\partial\psi}{\partial y} = 0.\end{aligned} \tag{12.7}$$

The coefficients of the derivatives in these equations are functions of the two dependent variables, $\overset{0}{\rho}$ and ψ; the equations are quasi-linear.

12.3.1 Examination of the Type of the Differential Equations

We compare the system (12.6) with the general system (7.1), (7.2) on page 77,

$$\begin{aligned}\overset{1}{A}_1\frac{\partial u}{\partial x}+\overset{1}{A}_2\frac{\partial u}{\partial y}+\overset{1}{B}_1\frac{\partial v}{\partial x}+\overset{1}{B}_2\frac{\partial v}{\partial y}&=\overset{1}{E}\\ \overset{2}{A}_1\frac{\partial u}{\partial x}+\overset{2}{A}_2\frac{\partial u}{\partial y}+\overset{2}{B}_1\frac{\partial v}{\partial x}+\overset{2}{B}_2\frac{\partial v}{\partial y}&=\overset{2}{E},\end{aligned}\tag{12.8}$$

and obtain the expressions for the various coefficients. The coefficients $\overset{k}{A}_i, i,k=1,2$ are:

$$\begin{aligned}\overset{1}{A}_1&=1+\sin\phi\cos(2\psi) & \overset{1}{A}_2&=\sin\phi\sin(2\psi)\\ \overset{2}{A}_1&=\sin\phi\sin(2\psi) & \overset{2}{A}_2&=1-\sin\phi\cos(2\psi),\end{aligned}\tag{12.9}$$

and the coefficients $\overset{k}{B}_i,\ i,k=1,2$ are

$$\begin{aligned}\overset{1}{B}_1&=-2\overset{0}{\rho}\sin\phi\sin(2\psi) & \overset{1}{B}_2&=2\overset{0}{\rho}\sin\phi\cos(2\psi)\\ \overset{2}{B}_1&=\ \ 2\overset{0}{\rho}\sin\phi\cos(2\psi) & \overset{2}{B}_2&=2\overset{0}{\rho}\sin\phi\sin(2\psi),\end{aligned}\tag{12.10}$$

and $\overset{1}{E}$ and $\overset{2}{E}$ are zero,

$$\overset{1}{E}=\beta_x=0\quad \overset{2}{E}=\beta_y=0.\tag{12.11}$$

We compute the constants a,b, and c, given by (7.15) on page 79, to determine the type of partial differential equations:

$$\begin{aligned}a&=\overset{1}{A}_1\overset{1}{B}_2-\overset{1}{A}_2\overset{1}{B}_1\\ b&=\overset{1}{A}_1\overset{2}{B}_2-\overset{1}{A}_2\overset{2}{B}_1+\overset{2}{A}_1\overset{1}{B}_2-\overset{2}{A}_2\overset{1}{B}_1\\ c&=\overset{2}{A}_1\overset{2}{B}_2-\overset{2}{A}_2\overset{2}{B}_1.\end{aligned}\tag{12.12}$$

We find the following expressions for a,

$$\begin{aligned}a&=2\overset{0}{\rho}\sin\phi\cos(2\psi)+2\overset{0}{\rho}\sin^2\phi\cos^2(2\psi)+2\overset{0}{\rho}\sin^2\phi\sin^2(2\psi)\\ &=2\overset{0}{\rho}\sin\phi[\sin\phi+\cos(2\psi)],\end{aligned}\tag{12.13}$$

for b,

$$\begin{aligned}b&=2\overset{0}{\rho}\sin\phi\sin(2\psi)+2\overset{0}{\rho}\sin^2\phi\sin(2\psi)\cos(2\psi)-2\overset{0}{\rho}\sin^2\phi\sin(2\psi)\cos(2\psi)\\ &\quad+2\overset{0}{\rho}\sin^2\phi\sin(2\psi)\cos(2\psi)-2\overset{0}{\rho}\sin\phi\sin(2\psi)\cos(2\psi)+2\overset{0}{\rho}\sin\phi\sin(2\psi)\\ &=4\overset{0}{\rho}\sin\phi\sin(2\psi),\end{aligned}\tag{12.14}$$

and for c,

$$\begin{aligned} c &= 2\overset{0}{\rho}\sin^2\phi\sin^2(2\psi) - 2\overset{0}{\rho}\sin\phi\cos(2\psi) + 2\overset{0}{\rho}\sin^2\phi\cos^2(2\psi) \\ &= 2\overset{0}{\rho}\sin\phi[\sin\phi - \cos(2\psi)]. \end{aligned} \tag{12.15}$$

The sign of the discriminant $b^2 - 4ac$,

$$\begin{aligned} b^2 - 4ac &= 16\overset{0}{\rho}{}^2\sin^2\phi\sin^2(2\psi) - 16\overset{0}{\rho}{}^2\sin^2\phi[\sin^2\phi - \cos^2(2\psi)] \\ &= 16\overset{0}{\rho}{}^2\sin^2\phi\cos^2\phi > 0, \end{aligned} \tag{12.16}$$

determines the type of the equations. The discriminant is positive (the case $\phi = 0$ is covered by setting $\overset{0}{\rho}\sin\phi$ equal to a constant); the system of equations is hyperbolic.

12.3.2 The Characteristics

We examined the hyperbolic partial differential equation in Chapter 7. We write the differential equation along the characteristics according to (7.21) on page 80:

$$\begin{aligned} \frac{du}{ds_1} + \underset{1}{\mu}\frac{dv}{ds_1} &= 0 \\ \frac{du}{ds_2} + \underset{2}{\mu}\frac{dv}{ds_2} &= 0, \end{aligned} \tag{12.17}$$

where

$$\underset{k}{\mu} = \frac{\underset{k}{\lambda}\overset{1}{B_1} + \overset{2}{B_1}}{\underset{k}{\lambda}\overset{1}{A_1} + \overset{2}{A_1}} \qquad k = 1,2. \tag{12.18}$$

We obtain expressions for the constants $\underset{k}{\lambda}$ from

$$\underset{1}{\lambda}, \underset{2}{\lambda} = \frac{-b \pm \sqrt{b^2 - 4ac}}{2a}. \tag{12.19}$$

The characteristics are parallel to the vectors $\underset{k}{\lambda}\overset{k}{A_i} = \underset{k}{\lambda}\overset{1}{A_i} + \overset{2}{A_i}$ $(k = 1,2)$; if $\underset{k}{\alpha}$ represents the angle between the k-characteristic and the x-axis, then

$$\tan\underset{k}{\alpha} = \frac{\underset{k}{\lambda}\overset{1}{A_2} + \overset{2}{A_2}}{\underset{k}{\lambda}\overset{1}{A_1} + \overset{2}{A_1}} = \frac{\underset{k}{\lambda}\overset{1}{B_2} + \overset{2}{B_2}}{\underset{k}{\lambda}\overset{1}{B_1} + \overset{2}{B_1}} \qquad k = 1,2. \tag{12.20}$$

We have, for $\underset{k}{\lambda}$:

$$\underset{k}{\lambda} = \frac{-b \pm \sqrt{b^2 - 4ac}}{2a} = \frac{-4\overset{0}{\rho}\sin\phi\sin(2\psi) \pm \sqrt{16\overset{0}{\rho}{}^2\sin^2\phi\cos^2\phi}}{4\overset{0}{\rho}\sin\phi[\sin\phi + \cos(2\psi)]}, \tag{12.21}$$

and divide numerator and denominator by $4\overset{0}{\rho}\sin\phi$,

$$\underset{k}{\lambda} = -\frac{\sin(2\psi) \mp \cos\phi}{\cos(2\psi) + \sin\phi}. \tag{12.22}$$

We write the numerator and the denominator as a product, using that $\cos\phi = \sin(\pi/2 - \phi)$ and $\sin\phi = \cos(\pi/2 - \phi)$:

$$\underset{k}{\lambda} = -\frac{\sin(2\psi) \mp \sin(\pi/2 - \phi)}{\cos(2\psi) + \cos(\pi/2 - \phi)} = -\frac{2\sin(\psi \mp \pi/4 \pm \phi/2)\cos(\psi \pm \pi/4 \mp \phi/2)}{2\cos(\psi + \pi/4 - \phi/2)\cos(\psi - \pi/4 + \phi/2)}, \tag{12.23}$$

or

$$\underset{k}{\lambda} = -\tan(\psi \mp \pi/4 \pm \phi/2) = -\tan\underset{k}{\beta}, \tag{12.24}$$

where

$$\underset{k}{\beta} = \psi \mp \pi/4 \pm \phi/2. \tag{12.25}$$

12.3.3 Characteristic Directions

We find the directions of the characteristics, $\tan\underset{k}{\alpha}$, from (12.20) with (12.10),

$$\tan\underset{k}{\alpha} = \frac{2\underset{k}{\lambda}\rho_0\sin\phi\cos(2\psi) + 2\rho_0\sin\phi\sin(2\psi)}{-2\underset{k}{\lambda}\rho_0\sin\phi\sin(2\psi) + 2\rho_0\sin\phi\cos(2\psi)}, \tag{12.26}$$

divide both the numerator and the denominator by $2\overset{0}{\rho}\sin\phi$, and use (12.24) for $\underset{k}{\lambda}$:

$$\tan\underset{k}{\alpha} = \frac{\underset{k}{\lambda}\cos(2\psi) + \sin(2\psi)}{-\underset{k}{\lambda}\sin(2\psi) + \cos(2\psi)} = \frac{-\cos(2\psi)\sin\underset{k}{\beta} + \sin(2\psi)\cos\underset{k}{\beta}}{\sin(2\psi)\sin\underset{k}{\beta} + \cos(2\psi)\cos\underset{k}{\beta}} = \frac{\sin(2\psi - \underset{k}{\beta})}{\cos(2\psi - \underset{k}{\beta})}. \tag{12.27}$$

or, with expression (12.25) for $\underset{k}{\beta}$,

$$\tan\underset{k}{\alpha} = \tan\{2\psi - (\psi \mp \pi/4 \pm \phi/2)\}. \tag{12.28}$$

The angles between the characteristics and the x-axis are:

$$\underset{k}{\alpha} = \psi \pm \pi/4 \mp \phi/2. \tag{12.29}$$

12.3.4 Equations Along the Characteristics

We compute the factors $\underset{k}{\mu}$ from (12.18), considering the denominator first,

$$\begin{aligned}\underset{k}{\lambda}\overset{1}{A}_1+\overset{2}{A}_1 &= -\tan\underset{k}{\beta}-\tan\underset{k}{\beta}\sin\phi\cos(2\psi)+\sin\phi\sin(2\psi)\\ &= -\frac{1}{\cos\underset{k}{\beta}}\left\{\sin\underset{k}{\beta}+\sin\phi\left[\sin\underset{k}{\beta}\cos(2\psi)-\cos\underset{k}{\beta}\sin(2\psi)\right]\right\}\\ &= -\frac{1}{\cos\underset{k}{\beta}}\left[\sin\underset{k}{\beta}-\sin\phi\sin(2\psi-\underset{k}{\beta})\right].\end{aligned} \tag{12.30}$$

We find for $\underset{k}{\lambda}\overset{1}{B}_1+\overset{2}{B}_1$:

$$\begin{aligned}\underset{k}{\lambda}\overset{1}{B}_1+\overset{1}{B}_2 &= 2\tan\underset{k}{\beta}\overset{0}{\rho}\sin\phi\sin(2\psi)+2\overset{0}{\rho}\sin\phi\cos(2\psi)\\ &= \frac{2\overset{0}{\rho}\sin\phi}{\cos\underset{k}{\beta}}\left[\cos(2\psi)\cos\underset{k}{\beta}+\sin\underset{k}{\beta}\sin(2\psi)\right]\\ &= \frac{2\overset{0}{\rho}\sin\phi}{\cos\underset{k}{\beta}}\cos(2\psi-\underset{k}{\beta}).\end{aligned} \tag{12.31}$$

We use the latter two expressions in (12.18),

$$\underset{k}{\mu}=\frac{\underset{k}{\lambda}\overset{1}{B}_1+\overset{2}{B}_1}{\underset{k}{\lambda}\overset{1}{A}_1+\overset{2}{A}_1}=2\overset{0}{\rho}\sin\phi\frac{\cos(2\psi-\underset{k}{\beta})}{\sin(2\psi-\underset{k}{\beta})\sin\phi-\sin\underset{k}{\beta}}. \tag{12.32}$$

We examine the denominator of this expression separately for $k=1,2$,

$$\sin(2\psi-\underset{1}{\beta})\sin\phi-\sin\underset{1}{\beta}=\sin(2\psi-\underset{1}{\beta})\sin\phi-\sin(\underset{2}{\beta}-\pi/2+\phi), \tag{12.33}$$

where we used that $\underset{1}{\beta}-\underset{2}{\beta}=-\pi/2+\phi$. We expand the trailing term at the right-hand side of (12.33),

$$\begin{aligned}&\sin(2\psi-\underset{1}{\beta})\sin\phi-\sin\underset{1}{\beta}\\ &=\sin(2\psi-\underset{1}{\beta})\sin\phi-[\sin\underset{2}{\beta}\cos(\pi/2-\phi)-\sin(\pi/2-\phi)\cos\underset{2}{\beta}]\\ &=\sin(2\psi-\underset{1}{\beta})\sin\phi-\sin\underset{2}{\beta}\sin\phi+\cos\phi\cos\underset{2}{\beta}.\end{aligned} \tag{12.34}$$

Since $2\psi - \underset{1}{\beta} = \underset{2}{\beta}$, the first two terms cancel,

$$\sin(2\psi - \underset{1}{\beta})\sin\phi - \sin\underset{1}{\beta} = \cos\phi\cos\underset{2}{\beta} = \cos\phi\cos(2\psi - \underset{1}{\beta}). \tag{12.35}$$

We use this in the expression for $\underset{1}{\mu}$, (12.32),

$$\underset{1}{\mu} = 2\overset{0}{\rho}\tan\phi. \tag{12.36}$$

We obtain a similar result for $k = 2$, but with the opposite sign; the final expression for $\underset{k}{\mu}$ is:

$$\underset{k}{\mu} = \pm 2\overset{0}{\rho}\tan\phi. \tag{12.37}$$

We are now in a position to write differential equations along the characteristics, see (12.17),

$$\begin{aligned} \frac{d\overset{0}{\rho}}{ds_1} + 2\overset{0}{\rho}\tan\phi\frac{d\psi}{ds_1} &= 0 \\ \frac{d\overset{0}{\rho}}{ds_2} - 2\overset{0}{\rho}\tan\phi\frac{d\psi}{ds_2} &= 0. \end{aligned} \tag{12.38}$$

These are Kötter's equations. We pose: $\overset{0}{\rho}\sin\phi = c$, $\cos\phi = 1$, $\phi = 0$, to obtain the equations for the special case that $\phi = 0$:

$$\begin{aligned} \frac{d\overset{0}{\rho}}{ds_1} + 2c\frac{d\psi}{ds_1} &= 0 \\ \frac{d\overset{0}{\rho}}{ds_2} - 2c\frac{d\psi}{ds_2} &= 0. \end{aligned} \tag{12.39}$$

We divide both equations in (12.38) by $2\overset{0}{\rho}\tan\phi$:

$$\begin{aligned} \tfrac{1}{2}\cot\phi\frac{1}{\overset{0}{\rho}}\frac{d\overset{0}{\rho}}{ds_1} + \frac{d\psi}{ds_1} &= 0 \\ \tfrac{1}{2}\cot\phi\frac{1}{\overset{0}{\rho}}\frac{d\overset{0}{\rho}}{ds_2} - \frac{d\psi}{ds_2} &= 0, \end{aligned} \tag{12.40}$$

and introduce a new variable χ,

$$\chi = \tfrac{1}{2}\cot\phi\ln\overset{0}{\rho}, \tag{12.41}$$

and write $1/\overset{0}{\rho}\int d\overset{0}{\rho}$ as $d\ln\overset{0}{\rho}$, so that the equations simplify to:

$$\begin{aligned} \chi + \psi &= C_1 \qquad \text{along } s_1 \\ \chi - \psi &= C_2 \qquad \text{along } s_2, \end{aligned} \tag{12.42}$$

where C_1 and C_2 are constants.

12.3.5 Application: Prandtl's Wedge

The solution for the problem of impending failure of the granular material below a long strip footing is one of the few solutions that exist for the quasi-linear partial differential equation studied in this chapter. The problem is illustrated in Figure 12.1. The load is applied along the strip $\mathscr{P}\mathscr{Q}$ of width $2b$; the normal stress is q_u and the shear stress is zero. Figure 12.1 corresponds to $\phi = \pi/6 = 30°$.

The major principal direction along the boundary $\mathscr{P}\mathscr{Q}$ is $\psi = -\pi/2$, so that the angle between the s_1-characteristics and the x-axis is $\underset{1}{\alpha} = -\pi/2 + \pi/4 - \phi/2 = -(\pi/4 + \phi/2) = -60°$, see (12.29). The s_1-characteristics are drawn as blue lines, and the s_2-characteristics are red. The functions $\chi = \chi_2$ and $\psi = \psi_2$ are given along the boundary $\mathscr{P}\mathscr{Q}$:

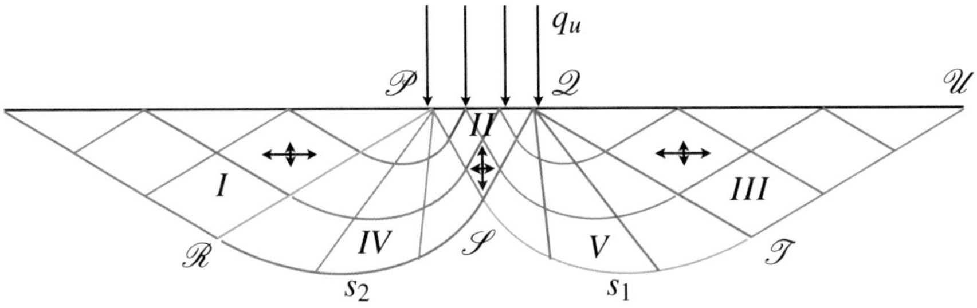

Figure 12.1: Prandtl's wedge with first (blue lines) and second (red lines) characteristics

$$\begin{aligned} \chi_2 &= \tfrac{1}{2}\cot\phi \ln \overset{0}{\rho}_2 = \tfrac{1}{2}\cot\phi \ln\left[\overset{0}{\sigma}_2 + c\cot\phi\right] & -b \le x \le b \quad y = 0 \\ \psi_2 &= -\pi/2 & -b \le x \le b \quad y = 0, \end{aligned} \tag{12.43}$$

where $\overset{0}{\sigma}_2$ is the value of the isotropic stress along the boundary. The isotropic stress equals the major principal stress, q_u, minus $\lambda = \overset{0}{\rho}_2 \sin\phi$ or

$$q_u - \overset{0}{\rho}_2 \sin\phi = \overset{0}{\sigma}_2 = \overset{0}{\rho}_2 - c\cot\phi, \tag{12.44}$$

so that

$$\overset{0}{\rho}_2 = \frac{q_u + c\cot\phi}{1 + \sin\phi}. \tag{12.45}$$

Intersecting an s_1-characteristic starting from the boundary $\mathscr{P}\mathscr{Q}$ with an s_2-characteristic, we find at the intersection point:

$$\begin{aligned} \chi + \psi &= \chi_2 - \pi/2 \\ \chi - \psi &= \chi_2 + \pi/2. \end{aligned} \tag{12.46}$$

Both equations hold at the point of intersection inside the triangle labeled II in Figure 12.1, so that the stresses are constant in that area; the characteristics are straight, as ψ is constant and equal to $-\pi/2$.

We proceed in the same way in zones I and III, where the major principal stress is horizontal, i.e., $\psi = 0$ in these triangles. The normal and shear stresses are zero along the boundary, so that the isotropic stress equals $\lambda = \overset{0}{\rho}_1 \sin\phi = \overset{0}{\rho}_3 \sin\phi$:

$$\overset{0}{\sigma}_3 = \overset{0}{\sigma}_1 = \lambda = \overset{0}{\rho}_3 - c\cot\phi = \overset{0}{\rho}_3 \sin\phi. \tag{12.47}$$

We solve for $\overset{0}{\rho}_3$

$$\overset{0}{\rho}_3 = \overset{0}{\rho}_1 = \frac{c\cot\phi}{1-\sin\phi}. \tag{12.48}$$

These values are constant in zones I and III, as follows from intersecting two characteristics of different kind starting at the boundary, and then solving the two equations (12.42).

The solution remains to be determined in the two fans $\mathscr{P}\mathscr{R}\mathscr{S}$ and $\mathscr{Q}\mathscr{S}\mathscr{T}$. We consider the fan $\mathscr{Q}\mathscr{S}\mathscr{T}$. The s_1-characteristics start from a single point, a singularity, at $\mathscr{Q}$, filling the space between the two bounding characteristics $\mathscr{Q}\mathscr{S}$ and $\mathscr{Q}\mathscr{T}$. We make the conjecture that these characteristics are straight lines, which implies that ψ is constant along them, and, by (12.42), $\overset{0}{\sigma}$ must be constant as well. The s_1-characteristics in the fan $\mathscr{Q}\mathscr{S}\mathscr{T}$ are curved, as shown in what follows. We find the values of χ and ψ along $\mathscr{Q}\mathscr{T}$: $\chi = \chi_3$, obtained from $\overset{0}{\sigma} = \overset{0}{\sigma}_3$, and $\psi = 0$. We have along $\mathscr{Q}\mathscr{S}$ that $\chi = \chi_2$, obtained from $\overset{0}{\sigma} = \overset{0}{\sigma}_2$, and $\psi = -\pi/2$. Since $\chi + \psi$ is constant along this characteristic, we have

$$\chi_2 + \psi_2 = \chi_3 + \psi_3. \tag{12.49}$$

We apply the definition of χ, (12.41),

$$\tfrac{1}{2}\cot\phi \ln \overset{0}{\rho}_2 - \pi/2 = \tfrac{1}{2}\cot\phi \ln \overset{0}{\rho}_3, \tag{12.50}$$

solve for $\ln \overset{0}{\rho}_2$

$$\ln \overset{0}{\rho}_2 = \ln \overset{0}{\rho}_3 + \pi \tan\phi, \tag{12.51}$$

and take the exponential on both sides:

$$\overset{0}{\rho}_2 = \overset{0}{\rho}_3 e^{\pi\tan\phi}. \tag{12.52}$$

We use this with (12.45) and (12.48) to obtain an equation for the limit load q_u,

$$\frac{q_u + c\cot\phi}{1+\sin\phi} = \frac{c\cot\phi}{1-\sin\phi} e^{\pi\tan\phi}, \tag{12.53}$$

so that

$$q_u = c\cot\phi \left[-1 + \frac{1+\sin\phi}{1-\sin\phi} e^{\pi\tan\phi} \right]. \tag{12.54}$$

This equation makes it possible to compute the maximum load that a strip footing can apply to a granular material of cohesion c and angle of internal friction ϕ.

12.3.6 Logarithmic Spirals

We determine equations for the s_1-characteristics in fan V based on the conjecture that the s_2-characteristics are straight. We apply the condition that the angles between the characteristics are constant and equal to $\pi/2-\phi$, see Figure 12.2. We use a polar coordinate system with the origin at $\mathscr{Q}$, and measure the angle θ from the x-axis, taking its sign negative; the domain is below the x-axis. The curves that result from this condition are logarithmic spirals; we obtain from the figure:

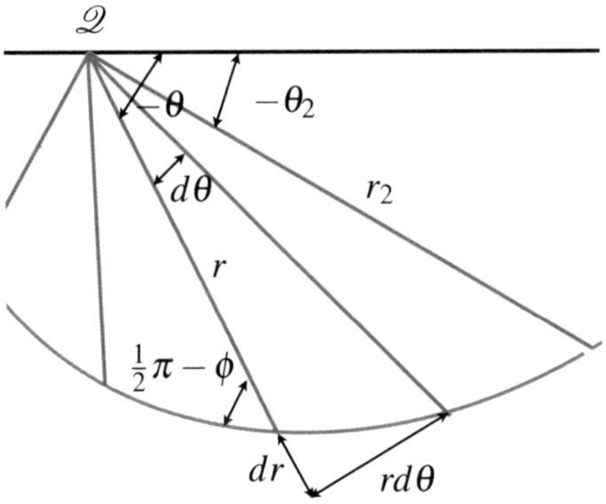

Figure 12.2: Logarithmic spiral, shown in blue

$$\frac{dr}{rd\theta}=\tan\phi, \tag{12.55}$$

and integrate,

$$\int_{\theta_1}^{\theta_2}\tan\phi d\theta=\int_{r_1}^{r_2}\frac{1}{r}dr, \tag{12.56}$$

so that

$$[\theta_2-\theta_1]\tan\phi=\ln r_2-\ln r_1=\ln(r_2/r_1). \tag{12.57}$$

We take the exponential of both sides of the equation,

$$r_2=r_1e^{(\theta_2-\theta_1)\tan\phi}. \tag{12.58}$$

This is the equation for the logarithmic spirals, compare with (9.57) on page 124.

12.3.7 Width of the Failure Zones Next to the Strip Load

Footing design requires that the width of the zones that may fail next to the strip load be known. We express the distance $\mathscr{Q}\mathscr{U}$ in terms of the footing width $B=2b$. The angle $\mathscr{S}\mathscr{Q}\mathscr{T}$ is $\pi/2$ (see Figure 12.3), $\theta_{\mathscr{Q}\mathscr{T}}-\theta_{\mathscr{Q}\mathscr{S}}=\pi/2$, and we express $r_{\mathscr{Q}\mathscr{T}}$ in terms of $r_{\mathscr{Q}\mathscr{S}}$ from (12.58),

$$r_{\mathscr{Q}\mathscr{T}}=r_{\mathscr{Q}\mathscr{S}}e^{(\theta_2-\theta_1)\tan\phi}=r_{\mathscr{Q}\mathscr{S}}e^{\pi/2\tan\phi}. \tag{12.59}$$

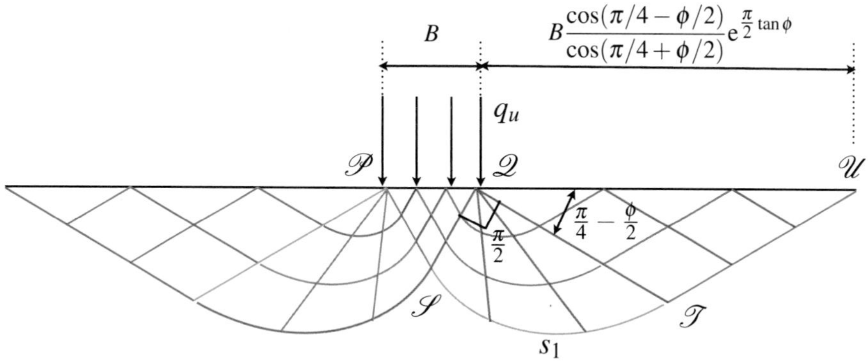

Figure 12.3: Computing the width of the zone of failure next to a strip load using the logarithmic spiral

We see from the figure that

$$\frac{\frac{1}{2}B}{r_{\mathscr{Q}\mathscr{S}}} = \cos\left(\frac{\pi}{4}+\frac{\phi}{2}\right), \tag{12.60}$$

and

$$\frac{\frac{1}{2}\overline{\mathscr{Q}\mathscr{U}}}{r_{\mathscr{Q}\mathscr{T}}} = \cos\left(\frac{\pi}{4}-\frac{\phi}{2}\right). \tag{12.61}$$

Hence,

$$\frac{\frac{1}{2}B}{r_{\mathscr{Q}\mathscr{S}}} * \frac{r_{\mathscr{Q}\mathscr{T}}}{\frac{1}{2}\overline{\mathscr{Q}\mathscr{U}}} = \frac{B}{\overline{\mathscr{Q}\mathscr{U}}} * \frac{r_{\mathscr{Q}\mathscr{T}}}{r_{\mathscr{Q}\mathscr{S}}} = \frac{\cos(\pi/4+\phi/2)}{\cos(\pi/4-\phi/2)}. \tag{12.62}$$

We apply (12.58) with $\theta_2-\theta_1=\pi/2$ to compute the ratio $r_{\mathscr{Q}\mathscr{T}}/r_{\mathscr{Q}\mathscr{S}}$,

$$\frac{B}{\overline{\mathscr{Q}\mathscr{U}}} * e^{\pi/2\tan\phi} = \frac{\cos(\pi/4+\phi/2)}{\cos(\pi/4-\phi/2)}, \tag{12.63}$$

or

$$\overline{\mathscr{Q}\mathscr{U}} = B\frac{\cos(\pi/4-\phi/2)}{\cos(\pi/4+\phi/2)}e^{\pi/2\tan\phi}. \tag{12.64}$$

Lowest Point of the Logarithmic Spiral

It is of practical value to determine the lowest point of the logarithmic spiral; it defines the greatest depth that the zone of impending failure reaches. We find the lowest point of the logarithmic spiral by setting θ to $-\pi/2+\phi$, see Figure 12.4, We follow the logarithmic spiral from the line $\mathscr{S}\mathscr{Q}$ in Figure 12.3, where $\theta=-3\pi/4-\phi/2$, to the lowest point, where $\theta=-\pi/2+\phi$, see Figure 12.4, so that the angle θ increases from $-3\pi/4+\phi/2$ to $-\pi/2+\phi$, i.e., an increment $\Delta\theta$ of $\pi/4+\phi/2$,

$$r = r_{QS}e^{(\pi/4+\phi/2)\tan\phi}. \tag{12.65}$$

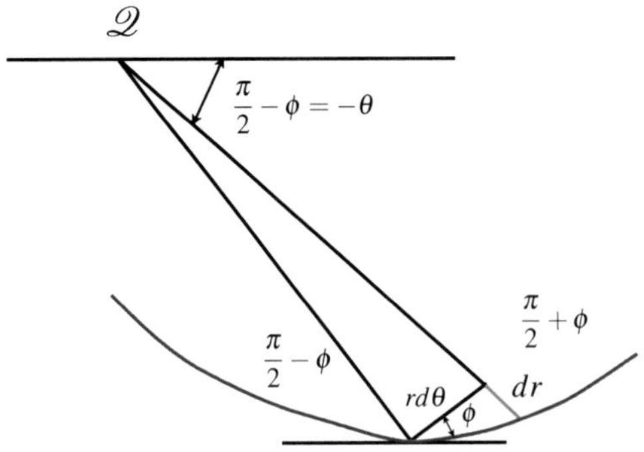

Figure 12.4: Determining the lowest point of the logarithmic spiral.

We obtain r_{QS} from (12.60), and compute $y_{\min}$ as $r\sin(\phi - \pi/2)$,

$$y_{\min} = \frac{Be^{(\pi/4+\phi/2)\tan\phi}}{2\cos(\pi/4+\phi/2)}\sin(\phi - \pi/2). \tag{12.66}$$

We simplify this equation by writing $\sin(\phi - \pi/2)$ as $2\sin(\phi/2 - \pi/4)\cos(\phi/2 - \pi/4)$

$$y_{\min} = B\frac{\sin(\phi/2 - \pi/4)\cos(\phi/2 - \pi/4)}{\cos(\pi/4+\phi/2)}e^{(\pi/4+\phi/2)\tan\phi} \tag{12.67}$$

We use that $\sin(\phi/2 - \pi/4) = -\cos(\phi/2 + \pi/4)$, and cancel the resulting common term in the numerator and denominator of the fraction:

$$y_{\min} = -B\cos(\pi/4 - \phi/2)e^{(\pi/4+\phi/2)\tan\phi}. \tag{12.68}$$

Chapter 13

The Navier-Stokes Equations

The Navier-Stokes equations are a set of highly non-linear partial differential equations. We present these equations as the final example of partial differential equations, because of their special character and their importance in the field of fluid mechanics. Common forms of these equations are obtained by simplification, for example by setting the viscosity of the fluid equal to zero (open channel flow), or by neglecting the acceleration relative to frictional losses, as in groundwater flow.

We consider fluid flow with the effect of viscosity included. In that case, the force F_i that acts on an elementary fluid volume involves the stresses, and (3.4) on page 18 now becomes, taking stresses positive for tension,

$$F_i = \partial_j \sigma_{ji} - \rho g \partial_i x_3. \tag{13.1}$$

We modify the form of Euler's equation given in (3.35) on page 24 to account for the stresses by replacing $-\partial_i p$ by $\partial_i \sigma_{ji}$ and obtain, representing the curl as C_k,

$$\frac{1}{g}\frac{\partial v_i}{\partial t} + \frac{1}{2g}\frac{\partial v^2}{\partial x_i} - \frac{1}{g}\varepsilon_{ijk} v_j C_k = \frac{1}{\rho g}\partial_j \sigma_{ji} - \frac{\partial x_3}{\partial x_i}. \tag{13.2}$$

We introduce the relationship for a Newtonian fluid between stresses and the velocity gradient $\partial_i v_j$. The general form, as originally proposed by Newton, is

$$\sigma_{ij} = -p\delta_{ij} + \lambda \delta_{ij}[\partial_p v_p] + \mu[\partial_i v_j + \partial_j v_i], \tag{13.3}$$

where μ [FT/L^2] is the viscosity.

We obtain an expression for the invariant σ_{ii} by setting i equal to j, with $\delta_{ii} = 3$

$$\sigma_{ii} = -3p + (3\lambda + 2\mu)\partial_i v_i. \tag{13.4}$$

We normally choose the factor λ such that

$$\sigma_{ii} = 3p, \tag{13.5}$$

which implies that

$$\lambda = -\tfrac{2}{3}\mu, \tag{13.6}$$

O. D. L. Strack, *Applications of Vector Analysis and Complex Variables in Engineering*, https://doi.org/10.1007/978-3-030-41168-8_13

so that (13.3) becomes,

$$\sigma_{ij} = -p\delta_{ij} - \tfrac{2}{3}\mu\delta_{ij}\partial_p v_p + \mu(\partial_j v_i + \partial_i v_j). \tag{13.7}$$

Equation (13.2) involves the derivatives $\partial_i\sigma_{ij}$, which we obtain from (13.7)

$$\partial_i\sigma_{ij} = -\partial_i p\delta_{ij} - \tfrac{2}{3}\mu\delta_{ij}\partial_i\partial_p v_p + \mu(\partial_i\partial_j v_i + \partial_i\partial_i v_j), \tag{13.8}$$

or, renaming the summed index i to p in the last term,

$$\partial_i\sigma_{ij} = -\partial_j p - \tfrac{2}{3}\mu\partial_j\partial_p v_p + \mu\partial_j\partial_p v_p + \mu\partial_p\partial_p v_j. \tag{13.9}$$

We collect terms and rewrite the trailing term,

$$\partial_i\sigma_{ij} = -\partial_j p + \tfrac{1}{3}\mu\partial_j\partial_p v_p + \mu\partial_p[\partial_p v_j - \partial_j v_p] + \mu\partial_j\partial_p v_p, \tag{13.10}$$

and modify the equation further,

$$\partial_i\sigma_{ij} = -\partial_j p + \tfrac{4}{3}\mu\partial_j\partial_p v_p + \mu\partial_p[\delta_{pm}\delta_{jn} - \delta_{pn}\delta_{jm}]\partial_m v_n, \tag{13.11}$$

or, finally,

$$\partial_i\sigma_{ij} = -\partial_j p + \tfrac{4}{3}\mu\partial_j\partial_p v_p + \mu\partial_p\varepsilon_{ipj}\varepsilon_{imn}\partial_m v_n. \tag{13.12}$$

We replace $\partial_p v_p$ by the divergence D, and $\varepsilon_{imn}\partial_m v_n$ by the curl C_i,

$$\partial_i\sigma_{ij} = -\partial_j p + \tfrac{4}{3}\mu\partial_j D + \mu\varepsilon_{ipj}\partial_p C_i. \tag{13.13}$$

The last term is minus the curl of C_k, and we rearrange indices,

$$\partial_i\sigma_{ij} = -\partial_j p + \tfrac{4}{3}\mu\partial_j D - \mu\varepsilon_{jpq}\partial_p C_q. \tag{13.14}$$

We substitute this expression for $\partial_i\sigma_{ij}$ in (13.2) and obtain, after renaming indices

$$\frac{1}{g}\frac{\partial v_i}{\partial t} + \frac{1}{2g}\frac{\partial v^2}{\partial x_i} - \frac{1}{g}\varepsilon_{ijk}v_j C_k = \frac{1}{\rho g}[-\partial_i p + \tfrac{4}{3}\mu\partial_i D - \mu\varepsilon_{ijk}\partial_j C_k] - \partial_i x_3. \tag{13.15}$$

We rearrange terms,

$$\partial_i\left[P + x_3 + \frac{v^2}{2g}\right] = -\frac{1}{g}\frac{\partial v_i}{\partial t} + \frac{1}{g}\varepsilon_{ijk}v_j C_k + \frac{\mu}{\rho g}\left[\tfrac{4}{3}\partial_i D - \varepsilon_{ijk}\partial_j C_k\right], \tag{13.16}$$

where P is defined in (3.36) on page 24 for the general case that the density varies.

Equations (13.16) are three equations, the Navier-Stokes equations. These non-linear partial differential equations are difficult to solve, but describe the general case of flow of a compressible viscous fluid. The unknowns are the three components of the velocity vector, the pressure p, and the density ρ. Together with the continuity equation (3.22) on page 21, and the relationship between pressure and density, they constitute the system of partial differential equations that fully describe the flow of Newtonian fluids.

13.1 Energy Transfer in a Fluid

We write $P + x_3 + v^2/(2g)$ as H, and use this in (13.16):

$$\partial_i H = -\frac{1}{g}\frac{\partial v_i}{\partial t} + \frac{1}{g}\varepsilon_{ijk} v_j C_k + \frac{\mu}{\rho g}[\tfrac{4}{3}\partial_i D - \varepsilon_{ijk}\partial_j C_k]. \tag{13.17}$$

For steady viscous flow, the energy head H is a constant if D and C_k are constant, i.e., if the flow is incompressible and irrotational; it remains true that energy dissipation, via mechanical energy, is possible only via either divergence or curl. We take the divergence of $\partial_i H$ and obtain,

$$\partial_i\partial_i H = \nabla^2 H = -\frac{1}{g}\frac{\partial[\partial_i v_i]}{\partial t} + C_k C_k + \frac{1}{g}\varepsilon_{ijk} v_j \partial_i C_k + \frac{\mu}{\rho g}[\tfrac{4}{3}\nabla^2 D - \varepsilon_{ijk}\partial_i\partial_j C_k]. \tag{13.18}$$

The last term vanishes, because of symmetry of the operator $\partial_i\partial_j$, and we obtain, with $\partial_i v_i = D$,

$$\nabla^2 H = -\frac{1}{g}\frac{\partial D}{\partial t} + C_k C_k - \frac{1}{g}\varepsilon_{ijk} v_i \partial_j C_k + \frac{4\mu}{3\rho g}\nabla^2 D. \tag{13.19}$$

If the flow is incompressible, the terms that contain the divergence D vanish,

$$\nabla^2 H = C_k C_k - \frac{1}{g}\varepsilon_{ijk} v_i \partial_j C_k. \tag{13.20}$$

This equation is independent of the viscosity μ, and is identical to the one we obtained for an in-viscid fluid, equation (3.49) on page 26.

If the flow is incompressible and irrotational, the mechanical energy is a harmonic function; the energy in the flow domain may be redistributed, but no net gain or loss of energy is possible. With or without viscosity, the only way for an incompressible fluid to gain or lose energy is via rotation.

13.1.1 Turbulence

The Navier-Stokes equations, combined with the continuity equation and the relationship between pressure and density, together constitute a system of equations that is sufficient to solve for the unknowns in the system: the pressure, the density of the fluid, and the three components of the velocity vector. The compressibility may be neglected for liquids, in which case the equations reduce to a system of four equations (continuity equation, $\partial_i v_i = 0$ and the three Navier-Stokes equations) for the four unknowns: the pressure p and the three components of the velocity vector v_i. This system of equations is not linear; the second term in (13.17) contains the cross product of the velocity vector and the curl of the velocity vector. Such equations do not guarantee unique solutions and discontinuities in the velocity field may appear. It is possible therefore that one given set of boundary conditions may support different solutions, which is why the transition between laminar flow and turbulent flow is not sharply defined and turbulent flow is unstable.

13.2 The Bernouilli Equation with Energy Losses; A Simplified Form of the Energy Equation

We consider equation (13.17) along a streamline, by forming the dot product of (13.17) with the unit vector T_i in the direction of flow, i.e.,

$$T_i = \frac{v_i}{v}, \tag{13.21}$$

and we obtain

$$T_i\partial_i H = -T_i\frac{1}{g}\frac{\partial v_i}{\partial t} + \frac{1}{g}\varepsilon_{ijk}\frac{v_i}{v}v_jC_k + \frac{\mu}{\rho g}[\tfrac{4}{3}T_i\partial_i D - \varepsilon_{ijk}T_i\partial_j C_k] \tag{13.22}$$

We use the following identity:

$$T_i\frac{\partial v_i}{\partial t} = T_i\frac{\partial (vT_i)}{\partial t} = vT_i\frac{\partial T_i}{\partial t} + \frac{\partial v}{\partial t} = \frac{\partial v}{\partial t}; \tag{13.23}$$

we used that the derivative of a unit vector is perpendicular to the unit vector, because the length of a unit vector cannot change; the dot product of T_i and $\partial T_i/\partial t$ is zero (this can be seen also from $\partial(T_iT_i)/\partial t = 2T_i\partial T_i/\partial t = 0$). The second term to the right of the equal sign in (13.22) vanishes and we obtain upon integration, noting that $T_i\partial_i = d/ds$

$$\int_{\overset{1}{x}_i}^{\overset{2}{x}_i} \frac{dH}{ds}ds = -\frac{1}{g}\int_{\overset{1}{x}_i}^{\overset{2}{x}_i} \frac{\partial v}{\partial t}ds + \int_{\overset{1}{x}_i}^{\overset{2}{x}_i} \frac{\mu}{\rho g}[\tfrac{4}{3}T_i\partial_i D - \varepsilon_{ijk}T_i\partial_j C_k]ds, \tag{13.24}$$

where the integration is from point $\overset{1}{x}_i$ on the streamline to point $\overset{2}{x}_i$ on the streamline. If the density is constant and the flow is incompressible, then the divergence D is zero,

$$H(\overset{2}{x}_i) - H(\overset{1}{x}_i) = -\frac{1}{g}\int_{\overset{1}{x}_i}^{\overset{2}{x}_i} \frac{\partial v}{\partial t}ds - \frac{\mu}{\rho g}\int_{\overset{1}{x}_i}^{\overset{2}{x}_i} \varepsilon_{ijk}T_i\partial_j C_k ds. \tag{13.25}$$

The first term to the right of the equal sign vanishes for steady flow, and the second one represents energy loss due to rotation.

Appendices

Appendix A

Numerical Integration of the Cauchy Integral

We consider numerical integration of the Cauchy integral given by (9.42) on page 121

$$\underset{j}{a_n} = \frac{1}{2\pi} \oint_{\mathscr{C}} \Omega(z(\delta))\mathrm{e}^{-in\theta} d\theta, \tag{A.1}$$

where Ω is holomorphic inside $\mathscr{C}$. We evaluate this integral by dividing the circle $\mathscr{C}$ into N segments, each of length $\Delta\theta$ and multiply the arc length of each segment by the value of the complex potential Ω at the center of the segment of arc. The radius of the circle in the $\underset{j}{Z}$ plane is 1 and the coordinate of the center of the arc number k is

$$\delta_k = \mathrm{e}^{\mathrm{i}\theta_k}. \tag{A.2}$$

We choose the first interval to span the arc between $\theta = -\frac{1}{2}\Delta\theta$ and $\theta = \frac{1}{2}\Delta\theta$, where

$$\Delta\theta = \frac{2\pi}{N} \tag{A.3}$$

The value of θ_k thus becomes

$$\theta_k = k\Delta\theta \qquad k = 1, 2 \cdots N. \tag{A.4}$$

The complex potential Ω requires a global value of its complex argument, i.e., of $z(\delta_k)$,

$$z_k = z(\delta_k) = z(\mathrm{e}^{\mathrm{i}k\Delta\theta}) = \underset{j}{z} + \underset{j}{Z}\underset{j}{R} = \underset{j}{z} + \mathrm{e}^{\mathrm{i}k\Delta\theta}\underset{j}{R}. \tag{A.5}$$

The integral reduces to the following sum:

$$\underset{j}{a_n} = \frac{1}{2\pi} \sum_{k=1}^{N} \Omega(z_k)\mathrm{e}^{-\mathrm{i}n\theta_k}\Delta\theta = \frac{\Delta\theta}{2\pi} \sum_{k=1}^{N} \Omega(z_k)\mathrm{e}^{-\mathrm{i}n\theta_k}, \tag{A.6}$$

O. D. L. Strack, *Applications of Vector Analysis and Complex Variables in Engineering*, https://doi.org/10.1007/978-3-030-41168-8

or, with (A.3)

$$\underset{j}{a_n} = \frac{1}{N} \sum_{k=1}^{N} \Omega(z_k) \mathrm{e}^{-in\theta_k}. \tag{A.7}$$

Efficiency of evaluation can be improved significantly by first computing an storing the values of Ω at the N points, and retrieving them as needed for each value of n, i.e., for each coefficient $\underset{j}{a_n}$.

Appendix B

List of Problems with Page Numbers

Chapter 1

Problem	Page number
1.1	6
1.2	7
1.3	8
1.4	9
1.5	9
1.6	9
1.7	9
1.8	9
1.9	9
1.10	9

Chapter 2

Problem	Page number
2.1	16
2.2	16
2.3	16
2.4	16
2.5	16
2.6	17

Chapter 3

Problem	Page number
3.1	31

© Springer Nature Switzerland AG 2020
O. D. L. Strack, *Applications of Vector Analysis and Complex Variables in Engineering*, https://doi.org/10.1007/978-3-030-41168-8

Chapter 5

Chapter 8

Chapter 9

Chapter 11

Bibliography

R. Barnes and I. Janković. Two-dimensional flow through large numbers of circular inhomogeneities. *J. Hydrol.*, 226:204–210, 1999.

H. S. Carslaw and J. C. Jaeger. *Conduction of heat in solids, 2nd Ed.* Oxford University Press, London, 1959.

R.V. Churchill. *Operational Mathematics*. McGraw-Hill Book Company, second edition, 1958.

J. Dupuit. *Études Théoriques et Pratiques sur le Mouvement des Eaux dans les Canaux Decouverts et à Travers les Terrains Perméables, 2´ieme ed.* Dunod, Paris, 1863.

A. Duschek and A. Hochrainer. *Grundzüge der Tensorrechnung in analytischer Darstellung: II. Tensoranalysis*. Springer–Verlag, Wien, Austria, third edition, 1970. 334 pp.

P. Forchheimer. Ueber die Ergiebigkeit von Brunnen- Anlagen und Sickerschlitzen. *Z. Architkt. Ing. Verlag*, 32:539–563, 1886.

A. E. Green and W. Zerna. *Theoretical Elasticity*. Dover Publications, Inc., New York, 1968. 457 pp.

A. A. Griffith. The phenomena of rupture and flow in solids. *Philosophical Transactions of the Royal Society of London*, 221:163–198, 1921. doi: 10.1098/rsta.1921.0006.

H. M. Haitjema. *Analytic Element Modeling of Groundwater Flow*. Academic Press, Inc., San Diego, U.S.A., 1995.

N. I. Muskhelishvili. *Some Basic Problems of the Mathematical Theory of Elasticity*. Noordhoff International Publishing, Leiden, The Netherlands, 1953.

R. Pennings. Dewatering below a clay layer. Private communication, 2018.

M. L. Salisbury and R. Barnes. Control equation formulation for circular inhomogeneities in the analytic element method. In *Proceedings of Analytic Element Modeling of Groundwater Flow*, pages 171–178, Indianapolis, Indiana, 1994.

R. F. Scott. *Principles of soil mechanics*. Addison-Wesley Publishing Company, 1963.

O. D. L. Strack, *Applications of Vector Analysis and Complex Variables in Engineering*, https://doi.org/10.1007/978-3-030-41168-8

I. S. Sokolnikoff. *Mathematical Theory of Elasticity*. Robert E. Krieger Publishing Company, Malabar, Florida, second edition, 1956. 476 pp.

V. V. Sokolovski. *Statics of Soil Media*. Butterworth Scientific, 1960.

O. D. L. Strack. *Groundwater Mechanics*. Prentice-Hall, Inc., Englewood Cliffs, New Jersey, 1989. 732 pp.

O. D. L. Strack. Theory and applications of the analytic element method. *Reviews of Geophysics*, 41(2):1–16, 2003.

O. D. L. Strack. *Analytical Groundwater Mechanics*. Cambridge University Press, University Printing House, Cambridge, CB28BS, United Kingdom, 1 edition, 2017. doi: 10.1017/9781316563144.

K. Terzaghi. *Erdbaumechanik*. Franz Deuticke, Vienna, 1925.

A. Verruijt. *An Introduction to Soils Mechanics*. Springer, 2017.

W. Wirtinger. Zur formalen Theorie der Funktionen von mehrenen komplexen Veranderlichen. *Mathematischen Annalen*, 97:357–375, 1927.

Index

O. D. L. Strack, *Applications of Vector Analysis and Complex Variables in Engineering*, https://doi.org/10.1007/978-3-030-41168-8

Printed in the United States
by Baker & Taylor Publisher Services